MW00509120

ANIMALS IN MOTION

AN ELECTRO-PHOTOGRAPHIC INVESTIGATION
OF CONSECUTIVE PHASES OF
ANIMAL PROGRESSIVE MOVEMENTS.

EADWEARD MUYBRIDGE

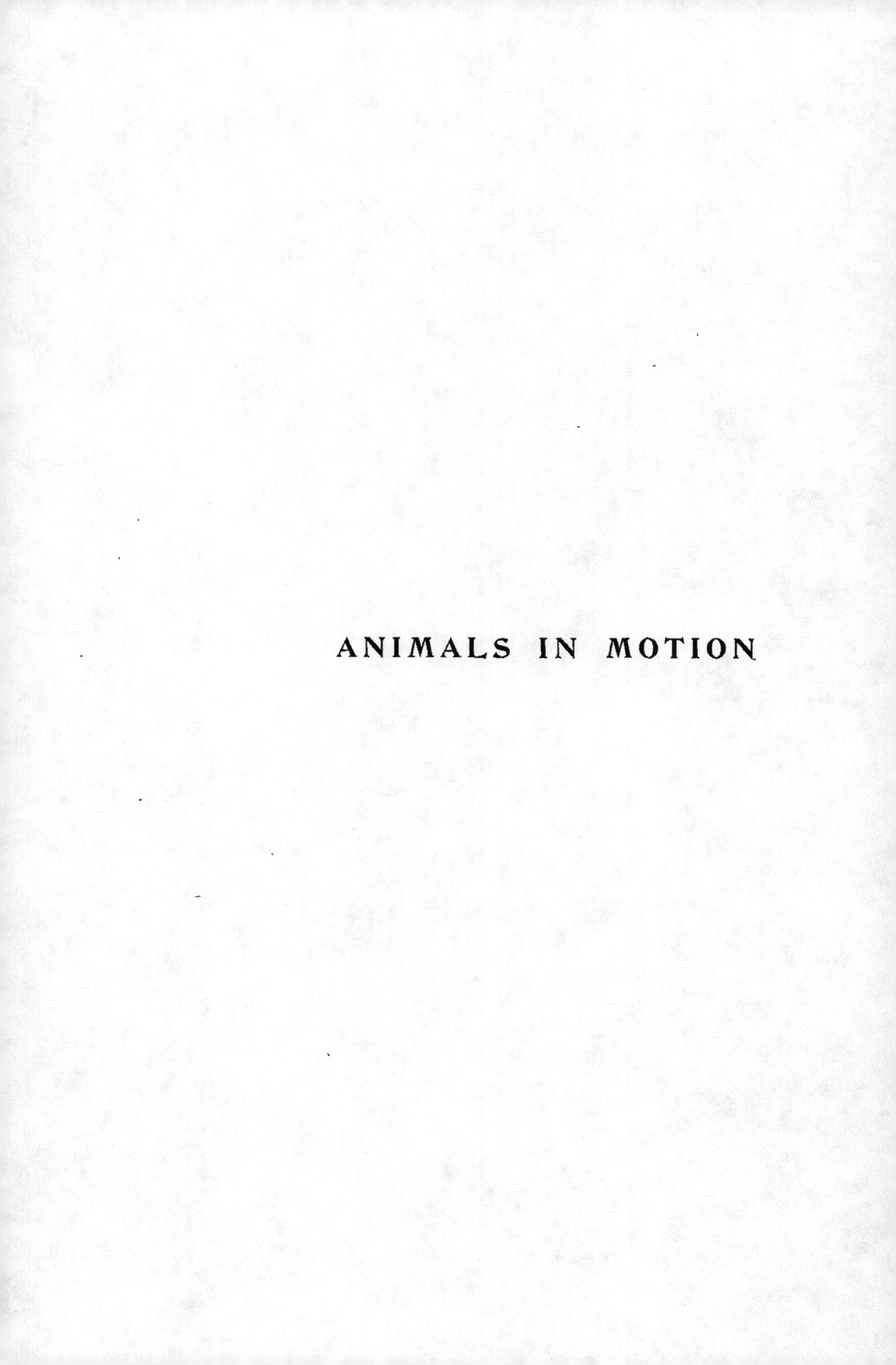

ANIMALS IN MOTION

From a photograph by the London Stereoscopic and Photographic Company.

Faithfully Yours

ANIMALS IN MOTION . . .

An Electro-Photographic Investigation of Consecutive Phases of Animal Progressive Movements . . . .

BY

EADWEARD MUYBRIDGE

Commenced 1872. Completed 1885

LONDON: CHAPMAN & HALL, LD.
. . . 1902 . . . .

PRINTED BY WILLIAM CLOWES AND SONS, LIMITED, LONDON AND BECCLES.

CONTENTS.

CATALOGUE OF ILLUSTRATIONS.

Description of Movement.	Animals.	Series.	Number of Phases. Laterals.	Fore-shortenings.	Total.	Page.
THE RACK	Text	....	....	...	....	133
	Horse ("Pronto") ...	41	12	12	24	137
	Egyptian Camel ...	42	16	....	16	139
	" " ...	....	3	....	3	141
	Horse ("Pronto") ...	....	....	3	3	141
THE CANTER	Text	....	1	...	1	143
	Horse ("Daisy") ...	43	23	...	23	147
	" ("Clinton") ...	44	12	....	12	149
	" " ...	....	6	....	6	151
	" (thorough-bred mare "Annie")	45	12	12	24	153
THE GALLOP	Text	....	13	4	17	155
Transverse-Gallop	Horse (thorough-bred "Bouquet") ...	46	22	....	22	171
	" (thorough-bred "Bouquet") ...	....	9	....	9	173
	" (thorough-bred mare "Annie")	47	16	...	16	175
	" (thorough-bred "Bouquet") ...	48	12	12	24	177
	" (thorough-bred "Bouquet") ...	49	12	12	24	179
	" (mare "Pandora") ...	50	....	24	24	181
	" ("Hansel") ...	51	....	24	24	181
	Buffalo (or Bison) ...	52	16	...	16	183
	Goat	53	24	....	24	185
	Bactrian Camel ...	54	20	....	20	187
	Cat	55	12	....	12	189
Rotatory-Gallop	Dog (racing-hound)	56	12	12	24	189
	" (mastiff) ...	57	8	8	16	191
	Fallow Deer	58	9	....	9	193
	Wapiti (or Elk) ...	59	20	....	20	195
	Fallow Deer	60	12	....	12	197
	Antelope	61	12	....	12	199
THE GALLOP Comparative	Cat, Bactrian Camel, Dog, Racoon, Wapiti, Bear, Goat, Buffalo, Hog	....	9	....	9	201
	Dogs, Cat, Buffalo, Goat, Wapiti, Bactrian Camel ...	....	6	3	9	203
THE RICOCHET	Text	...	—	—	—	205
	Kangaroo ...	62–3	13	—	13	207
THE LEAP	Text	—	3	—	3	209
	Horse (mare "Pandora")	64	20	—	20	211
	" "	65	—	16	16	213
	" "	66	12	—	12	215
	" "	67	—	24	24	215
	" ("Daisy") ...	68	20	—	20	217
	" "	69	20	—	20	219
	" (mare "Pandora")	—	6	—	6	221
	" "	—	—	12	12	223
	Cat ...	—	12	—	12	225
THE BUCK AND THE KICK	Text	—	—	—	—	227
	Mule ("Ruth") ...	70	8	—	8	229
	" " ...	71	8	—	8	229
	" "	—	7	—	7	231
CHANGE OF GAIT	Text	—	....	—	—	233
	Horse, walk to trot ...	72	24	—	24	235
	" trot to gallop	73	24	—	24	235
	" rack to gallop	74	20	—	20	235
	" recovery from a leap ...	75	24	—	24	235
THE FLIGHT OF BIRDS	Text	—	1	—	1	237
	Pigeon ...	76	12	12	24	239
	" ...	77	12	12	24	241
	Vulture ...	78	18	—	18	243
	Sulphur-crested cockatoo	79	24	—	24	245
	"	80	24	—	24	247
	"	81	—	20	20	249
	"	—	6	—	6	251
	"	—	—	7	7	253
THE FLYING RUN	Ostrich ...	82	12	12	24	255

PREFACE.

In the spring of the year 1872, while the author was directing the photographic surveys of the United States Government on the Pacific Coast, there was revived in the city of San Francisco a controversy in regard to animal locomotion, which we may infer, on the authority of Plato, was warmly argued by the ancient Egyptians, and which probably had its origin in the studio of the primitive artist when he submitted to a group of critical friends his first etching of a mammoth crushing through the forest, or of a reindeer grazing on the plains.

In this modern instance, the principal subject of dispute was the possibility of a horse, while trotting—even at the height of his speed—having all four of his feet, at any portion of his stride, simultaneously free from contact with the ground.

The attention of the author was directed to this controversy, and he immediately resolved to attempt its settlement.

The problem before him was, to obtain a sufficiently well-developed and contrasted image on a wet collodion plate, after an exposure of so brief a duration that a horse's foot, moving with a velocity of more than thirty yards in a second of time, should be photographed with its outlines practically sharp.

In those days the rapid dry process—by the use of which such an operation is now easily accomplished—had not been discovered. Every photographer was, in a great measure, his own chemist; he prepared his own dipping baths, made his own collodion, coated and developed his own plates, and frequently manufactured the chemicals necessary for his work. All this involved a vast amount of tedious and careful manipulation from which the present generation is, happily, relieved.

Having constructed some special exposing apparatus, and bestowed more than usual care in the preparation of the materials he was accustomed to use for ordinarily quick work, the author commenced his investigation on the race-track at Sacramento, California, in May, 1872, where he in a few days made several negatives of a celebrated horse, named Occident, while trotting, laterally, in front of his camera, at rates of speed varying from two minutes and twenty-five seconds to two minutes and eighteen seconds per mile.

The photographs resulting from this experiment were sufficiently sharp to give a recognizable silhouette portrait of the driver, and some of them exhibited the horse with all four of his feet clearly lifted, at the same time, above the surface of the ground.

So far as the immediate point at issue was concerned, the object of the experiment was accomplished, and the question settled for once and for all time in favour of those who argued for a period of unsupported transit.

Each of the photographs made at this time illustrated a more or less different phase of the trotting action. Selecting a number of these, the author endeavoured to arrange the consecutive phases of a complete stride; this, however, in consequence of the irregularity of their intervals, he was unable to satisfactorily accomplish.

It then occurred to him that a series of photographic images made in rapid succession at properly regulated intervals of time, or of distance, would definitely set at rest the many existing theories and conflicting opinions upon animal movements generally.

Having submitted his plans to Mr. Leland Stanford, who owned a number of thorough-breds, and first-class trotting horses, the author secured that gentleman's co-operation for a continuance of the researches at his stock-farm—now the site of the University—at Palo Alto.

His official and other duties, requiring absences from the city on expeditions sometimes extending over several months at a time, prevented continuous attention to the investigation, but in the meanwhile he devised a system for obtaining a succession of automatic exposures at intervals of time, which could be regulated at discretion.

The apparatus used for this initiatory work included a motor-clock for making and breaking electric circuits, which is briefly described in the "Proceedings of the Royal Institution of Great Britain," March 13, 1882, and will be, with the other arrangements, explained in detail further on.

Experiments were carried on from time to time as opportunity permitted; they were, however, principally for private or personal use, and it was not until 1878 that the results of any of them were published.

In that year the author deposited in the Library of Congress at Washington a number of sheets of photographs, each one of which illustrated several equidistant consecutive phases of one complete stride of a horse while walking, trotting, galloping, and so forth; they were published with the general title of "The Horse in Motion."

Some of these photographs found their way to distant parts of the world, and were reproduced and commented upon in the *Scientific American* (New York), October 19, 1878; *La Nature* (Paris), December 14, 1878; *Berliner Fremdenblatt* (Berlin), April 26, 1879; *Wiener Landwirthschaftliche Zeitung* (Vienna), April 26, 1879; *The Field* (London), June 28, 1879, and many other journals and magazines.

Each of the cameras used at this time had two lenses,

and made stereoscopic pictures. Selecting from these stereographs a suitable number of phases to reconstitute a full stride, he placed the appropriate halves of each, respectively, in one of the scientific toys called the zoetrope, or the wheel of life—an instrument originated by the Belgian physicist Plateau, to demonstrate the persistency of vision. These two zoetropes were geared, and caused to revolve at the same rate of speed; the respective halves of the stereographs were made simultaneously visible, by means of mirrors—arranged on the principle of Wheatstone's reflecting stereoscope—successively and intermittently, through the perforations in the cylinders of the instruments, with the result of a very satisfactory reproduction of an apparently solid miniature horse trotting, and of another galloping.

Pursuing this scheme, the author arranged, in the same consecutive order, on some glass discs, a number of equidistant phases of certain movements; each series, as before, illustrated one or more complete and recurring acts of motion, or a combination of them: for example, an athlete turning a somersault on horseback, while the animal was cantering; a horse making a few strides of the gallop, a leap over a hurdle, another few strides, another leap, and so on; or a group of galloping horses.

Suitable gearing of an apparatus constructed for the purpose caused one of these glass discs, when attached to a central shaft, to revolve in front of the condensing lens of a projecting lantern, parallel with, and close to another disc fixed to a tubular shaft which encircled the other, and around which it rotated in the contrary direction. The latter disc was of sheet-metal, in which, near its periphery, radiating from its centre, were long narrow perforations, the number of which had a definite relation to the number of phases in the one or more lines of motion on the glass disc—the same number, one or two more, or one or two less—according to the sequence of phases, the intended direction of the movement, or the variations desired in the apparent rate of speed.

The discs being of large size, small portions only of their surfaces—showing one phase of each of the circles of moving animals—were in front of the condenser at the same instant.

To correct the apparent vertical extension of the animals when seen through the narrow openings of the metal disc on its revolution in such close proximity to, and in the reverse direction of the glass disc, the photographs on the latter, after numerous experiments, were ultimately prepared as follows:—

A flexible positive was conically bent inwards, and inclined at the necessary angle from the lens of the copying camera to ensure the required horizontal elongation of the animal while the straight line of ground corresponded with the curvature of the intended ground-line of the glass disc, towards the periphery of which the feet of the animals were always pointed.

A negative was then made of this phase, and negatives of the other phases, in the same manner. All the negatives required for that particular subject were then consecutively arranged, equidistantly, in a circle, on a large sheet of glass; if the disc was to include more than one subject, the phases thereof were arranged in the same manner, and a transparent positive made of them collectively. The glass support of the resulting positive was subsequently cut into the form of a circle, and a hole bored through the centre, for the purpose of attaching it to the inner shaft of the apparatus.

Some of the discs illustrated eight or ten distinct seriates of 17, 18, 19, 20, or 21 phases each, arranged, with due regard to perspective effect, on different lines, and included perhaps 200 figures of animals, which successively appeared, the size of life, on the screen as if trotting, cantering, galloping, and so forth, in various directions, and at different rates of speed, and of men performing acts of non-progressive movement, such as bowing, or waving their arms. These apparent movements could be continued for a period limited only by the patience of the spectators. Much time and care were required in the preparation of the discs, each figure having to be photographed three times, independently, before being photographed collectively

For many of the discs it was found advisable to fill up the outlines with opaque paint, as a more convenient and satisfactory method of obtaining greater brilliancy and stronger contrasts on the screen than was possible with chemical manipulation only. In the "retouching" great care was invariably taken to preserve the photographic outline intact.

To this instrument the author gave the name of ZOÖPRAXISCOPE; it is the first apparatus ever used, or constructed, for synthetically demonstrating movements analytically photographed from life, and in its resulting effects is the prototype of all the various instruments which, under a variety of names, are used for a similar purpose at the present day.

In an article—"Photographs of a Galloping Horse"—published in the *Gentleman's Magazine*, December, 1881, Proctor, the astronomer, writes of having seen the zoopraxiscope in operation "about two years" before that date. This occurred at a lecture by the author on Animal Movements, given to the San Francisco Art Association.

The first demonstration given in Europe with the zoopraxiscope was at the laboratory of Dr. E. J. Marey, in the presence of a large number of scientists from various parts of the world, then attending the Electrical Congress at Paris. A detailed criticism thereof appeared in *Le Globe*, and other Parisian journals, September 27, 1881. The same apparatus was used at a lecture given by the author at the Royal Institution of Great Britain, the Prince of Wales presiding on the occasion; a long description (written by G. A. Sala) of the realistic effects of the synthetic projections then made, appeared in the *Illustrated London News*, March 18, 1882.

It may here be parenthetically remarked that on the 27th of February, 1888, the author, having contemplated some improvements of the zoöpraxiscope, consulted with Mr. Thomas A. Edison as to the practicability of using that instrument in association with the phonograph, so as to combine, and reproduce simultaneously, in the presence of an audience, visible actions and audible words. At that time the phonograph had not been adapted to reach the ears of a large audience, so the scheme was temporarily abandoned.

Five years after this interview, or twelve years after the zoopraxiscope had been exhibited at a large number of scientific and artistic institutions in Europe and America, the first improvement thereof, for the purpose of realizing the same effects, appeared in the instrument called by its ingenious constructor the "kinetoscope." This improvement was made possible by the invention of the celluloid ribbon, by the use of which a larger number of successive phases of motion could be obtained in the making of the original negatives than on glass plates, and in the synthetic exhibition of the positives thereof, than was possible on a glass disc, however large, or however close together the successive phases could be spirally arranged.

A great many claimants have arisen for this improvement on the zoopraxiscope. To Marey must be attributed the first obtainment, with a single lens, on a strip of sensitized material, of a succession of moving figures, which he accomplished in 1882; and to Edison the first application of a strip or ribbon containing a number of such figures in a straight line (instead of being arranged on a large glass disc), for lantern projection; this, after much patient attention, he succeeded in doing in 1893.

The combination of such an instrument with the phonograph has not, at the time of writing, been satisfactorily accomplished; there can, however, be but little doubt that in the — perhaps not far distant — future instruments will be constructed that will not only reproduce visible actions simultaneously with audible words, but an entire opera, with the gestures, facial expressions, and songs of the performers, with all the accompanying music, will be recorded and reproduced by an apparatus, combining the principles of the zoopraxiscope and the phonograph, for the instruction or entertainment of an audience long after the original participants shall have passed away; and if the photographs should have been made stereoscopically, and projections from each series be independently and synchronously projected on a screen, a perfectly realistic imitation of the original performance will be seen, in the apparent "round," by the use of properly constructed binocular glasses.

With the exception of a series of phases of a solar eclipse, made in January, 1880, the Palo Alto researches were concluded in 1879, they resulted in the publication of a volume containing about two hundred sheets of photographs, the greater number of which illustrated from twelve to twenty-four consecutive phases of some act of movement by a man, a horse, or some other animal. A few photographs were made of birds on the wing, others of groups of horses on the gallop, and many represented phases synchronously photographed from five different points of view.

A number of these subjects were, a few years afterwards, copied, and republished in a book bearing the same title as that originally used by their author, without the formality of placing his name on the title-page.

Although these preliminary labours had completely

demonstrated all the mysteries connected with the various gaits of a horse, it was recognized that the work was incomplete, in consequence of the difficulty, and sometimes impossibility, of obtaining, with wet collodion plates, the essential details of rapid muscular action. The results, however, excited so much attention in the artistic and scientific worlds, that the author was convinced a more systematic and comprehensive investigation, with the use of the then newly discovered dry-plate process, would result in the disclosure of a vast deal of information valuable alike to the artist and to the scientist, and of interest to the public generally.

The cost of an investigation on the contemplated scale, and the subsequent publication of the results in a commensurate manner, assumed such imposing proportions, that all publishers to whom the proposition was made shrunk—perhaps not unnaturally—from entering a field so fraught with possibilities of unremunerative outlay.

It was under these circumstances that the University of Pennsylvania—through the influence of its Provost, Dr. William Pepper—with an enlightened exercise of its functions as a contributor to human knowledge, instructed the author to make, under its auspices, a new and comprehensive investigation of animal movements, in the broadest signification of the words, and some of the trustees and friends of the university constituted themselves a committee for the purpose of promoting the execution of the work.

These gentlemen were Dr. William Pepper, Charles C. Harrison, Edward H. Coates, Samuel Dickson, J. B. Lippincott, and Thomas Hockley. It is with much gratification the author acknowledges his indebtedness to these gentlemen for the interest they took in his labours, for without their generous assistance the work would probably never have been undertaken. Among many others who also rendered valuable aid in his researches were Doctors F. X. Dercum, Geo. F. Barker, Horace H. Furness, Horace Jayne, and S. Weir Mitchell, of the university; Craige Lippincott, Arthur E. Brown (Director of the Zoological Gardens), and Lino F. Rondinella and Henry Bell, the author's two chief assistants, who, respectively, had charge of the electrical and developing departments.

The outdoor labours were recommenced in the spring of 1884, and terminated in the autumn of 1885. More than a hundred thousand photographic plates were used in the preparation of the work for the press. The results were published in 1887, with the title of "Animal Locomotion." The work contains more than 20,000 figures of moving men, women, children, beasts, and birds, in 781 photo-engravings, bound in eleven folio volumes.

The great cost of printing and manufacturing the work—independently of the preliminary expenses—necessarily restricted its sale to a comparatively few complete copies.

With a view of supplying the demand of art and of science students, and others whom the subject interests, it has been decided to select a number of the most important plates made at the university, and to republish them on a reduced scale in a more popular and accessible manner.

These plates demonstrate certain facts which occur, in regular sequence, with uniform intervals of time, during the accomplishment of some act of motion, thus enabling the phases which characterize the transition from one period of a movement to another period to be leisurely studied. They are chemically executed engravings, and are reproduced with all the original defects of photographic manipulation, precisely as they were made in the camera. A few plates of the Palo Alto investigation of 1872–79 are also included.

It is the hope of the author that the selection of subjects has been judiciously made, and that the artist will make discreet use of them. Should certain phases of the movements be considered of sufficient naturally artistic value to permit their being copied without derogation to artistic effect, it is unnecessary to say it is not for that purpose they are published; their mission is simply to furnish a guide to the laws which control animal movements, and to show how those movements are effected.

When the student has carefully noted the consecutive phases of some act of progressive motion, he will do wisely to put the book on one side and to seek his impression of that motion from the animal itself.

By this process of study he will unquestionably recognize the differences between his own educated impression of any of the analyzed gaits—especially those of the walk and of the gallop—and the interpretation given to them by one whose judgment is founded on tradition or unassisted observation; and he will be convinced that the concrete facts of animal locomotion can be ideally reproduced without offence to the canons of Art or sacrifice of the truth of Nature.

E. M.

KINGSTON-ON-THAMES,
December, 1898.

INTRODUCTION.

ZOOPRAXOGRAPHY, or the science of animal motion, has been studied by mankind from the most remote period of the world's history.

If we seek for evidence of its original application to design in art, we must direct our attention to an epoch very much nearer to that in which intelligent life first appeared on this earth, than to any of which history or even tradition has left the faintest connecting record.

Until the middle of this century it was the general belief that the most ancient relics of the attention of man to artistic pursuits would be found on the banks of the Nile.

The dates of Chinese antiquities were shrouded in mystery; all traces of the once powerful Hittites had disappeared, and the great cities of the Chaldæan and of the Assyrian empires had been so completely obliterated that Xenophon, two thousand years ago, marched his army over the site of Nineveh without apparent knowledge of its buried ruins.

About fifty years ago some explorations in the south of France brought to light a few remnants of carving and engraving executed by a race of men who, unknown centuries ago, left evidences of their sharing with the mammoth and the reindeer a life amid such circumstances as are experienced in an arctic region.

Living under the conditions which must have surrounded man at this early period of his evolution, it seems astonishing that he should have had either the inclination or the taste to devote his attention to artistic pursuits; but the debris of the caves wherein he dwelt furnish the proof not only of his being a skilful imitator, both in the round and in outline, of things which he saw, but, what is of especial interest, that art was born in his attempt to delineate an animal in motion. The few discovered fragments of his labours evince a quickness of observation, an appreciation of form and proportion, and a faculty of expressing movement with such scientific fidelity that as little imagination is required to understand the intention of the artist as he himself uses in the execution of his work.

The art of expressing ideas, or conveying information by pictorial representation, naturally long preceded the invention of letters; these etchings and carvings of animals by primitive man were apparently executed for no other purpose than to gratify his artistic impulses, or to record an event for which the language of his time was inadequate.

The importance of a correct knowledge of the various functions of the limbs in locomotion was recognized in the early ages of Greece. Aristotle devoted much attention to the subject, and Xenophon, Pliny, Vegetius, and many other writers of ancient times tell us of the great interest manifested by their respective contemporaries in the training and employments of the horse.

None of these distinguished authors, however, have left us any information as to the manner in which the various movements they write about were executed.

No investigation—worthy of the name—is known to have been made until the middle of the seventeenth century.

In 1680 Borelli of Naples published his "De Motu Animalium." This celebrated mathematician evidently conducted his experiments with great care; but, although he disapproved of the generally accepted idea of the walk, he allowed the Egyptian interpretation of the gallop to pass unchallenged.

In 1658 the Marquis of Newcastle published at Antwerp his elaborate and well-known work on "Horsemanship." The original was in the French language. A complete English edition was issued in 1743. The two folio volumes were illustrated with many finely executed copperplate engravings of horses performing various feats of motion, and a chapter was devoted to "The Movements of a Horse in all his Natural Paces."

As these analyses were the sources from which all lexicographers—English, French, and German—from the date of publication to the end of the last decade, seem to have taken their definitions of the various gaits of a horse, they are included in an appendix to this volume.

In the early part of this century, Ernst and Wilhelm Weber, the former an eminent physiologist, the latter an equally distinguished physicist, published at Leipzig the results of many years' patient devotion to the subject. The researches of Dr. E. J. Marey, of the College of France, and Dr. J. Bell Pettigrew, of the University of St. Andrews, are of too recent date, and are too well known, to require more than a passing allusion. The former was the first to avail himself of scientific appliances to automatically register the characteristics of movements. This was done by an ingenious apparatus, carried by the rider of an animal, which caused styles—actuated by pneumatic pressure—to leave a record on a revolving cylinder.

Although the actinic qualities of light, when its rays were directed to organic matter which had been treated with certain chemicals—especially salts of silver—was well known to the alchemists of the Middle Ages, and is now in universal use as a method of picture-making; it was only about a quarter of a century ago that its value as a factor in scientific research was beginning to be recognized. Since then it has been the means of opening many new fields for inquiry, and a number of important discoveries have been made which without its aid would have been impossible.

It was to photography, therefore, that the author resorted when he commenced a solution of the complex problem of "Animal Locomotion." The origin of his labours has been mentioned in the Preface. Before attempting a description of the results, it is essential that the system employed for their obtainment should be understood. The investigation at Palo Alto was conducted in practically the same manner as that at the University of Pennsylvania; it will, therefore, be sufficient to give a general explanation of the studio arrangements at the latter place.

Diagram of the Studio at the University of Pennsylvania, and Arrangement of the Apparatus for investigating Animal Locomotion.

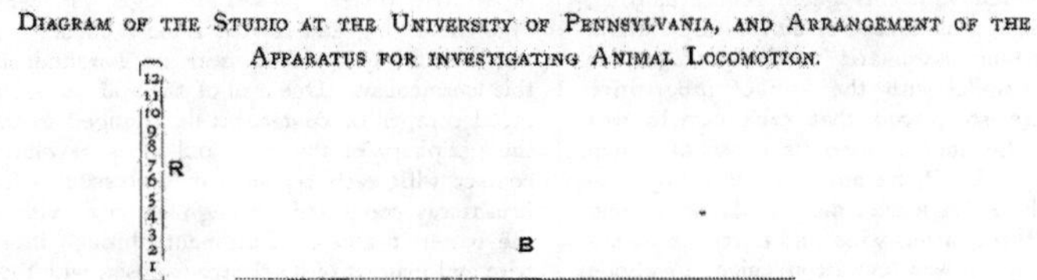

B, the lateral background, divided by threads into spaces five centimetres, or about two inches, square, every tenth thread being of greater thickness than the others. At C C were portable backgrounds divided into similar squares, and placed close to and at right angles with the lateral background, when a fore-shortened series of exposures rendered them necessary. These backgrounds were black or white, as circumstances required. T T, the track, covered with corrugated rubber matting, along which the animal was caused to move. L, a lateral battery of twenty-four automatic electro-photographic cameras, arranged parallel with the line of progressive motion; these were so placed that each camera was practically opposite the animal when its phase of movement was photographed. R, an automatic electro-photographic camera with twelve lenses, the plate-holders being adapted for a plate three inches wide and thirty-six inches long, which, in practice, it was found convenient to divide into three parts of twelve inches each. A supplementary lens was arranged to permit focussing while the plate-holder was in position. For fore-shortenings at an angle of ninety degrees to the laterals, this camera was usually placed on end, so as to obtain one vertical line of view for the series of twelve exposures. F, a similar camera to R, usually placed horizontally at any convenient point. D, the station of the director, and where were also placed the electric batteries, the motor-clock for intermittently completing the electric circuits, the chronograph for recording the intervals of time between each successive exposure, and other apparatus connected with the work.

The motive power of the circuit-maker was an adjustable weight attached to a cord wound round a drum, the speed being regulated, partially, by a fan-wheel.

Fastened to the frame of the motor-clock was a ring of hard rubber, in which were inserted twenty-four insulated segments of platinum; these segments were connected by insulated wires to the same number of binding-posts. A shaft, connected by an arrangement of geared wheels, passed through the centre of the segmented ring and carried a loose collar; a stout metal rod was firmly attached near its longitudinal centre to this loose collar. One arm of the rod carried a laminated metal scraper, or contact-brush, arranged to travel around the periphery of the ring, and in its revolution to make contact with each segment in succession. The contact-brush was connected, through its arm, with one pole of the battery; and each segment, through its independent wire and magnet of its electro-exposer, with the other pole.

When twenty-four consecutive phases of an act of motion were to be photographed, all of the insulated segments of the ring were put in circuit. When twelve consecutive phases were to be made synchronously from each of two or three points of view, each alternate segment was placed in circuit with the electric battery, and the proper connections made with each set of exposers in front of the lenses.

An experimental trial having been made to ascertain the time required by the animal or model to complete the intended movement, the weights and fan-wheel were adjusted to cause the contact-brush to sweep around the periphery of the segmented ring at the required rate of speed, and the apparatus was set in operation.

If the series was to illustrate progressive motion, the model or animal would start on its journey at a more or less distant point from the cameras. On its approach to the figure 1 on the track, and as near thereto as its speed and the personal equation of the director allowed, an independent current was switched on through a magnet below the rod of the contact-brush.

The action of the armature released the lower end of the rod on the loose collar, which, by means of a coiled spring, was immediately thrown into gearing with the already revolving shaft; the contact-brush swept around the segmented ring, and made the consecutive series of exposures at the pre-arranged intervals of time.

Providing it was so intended, three exposures were made simultaneously, one through each of the lenses marked 1 at each of the operating points marked respectively L, R, and F. At the pre-determined interval of time, another three synchronous exposures were made through each of the lenses marked 2, and so on until the entire series was completed. If the time had been accurately calculated, the successive phases were photographed in precise accordance with the arrival of the model at the numbered places on the track, and exactly opposite the correspondingly numbered camera. This perfect uniformity of time, speed, and distance was not always obtained, and allowance was therefore usually made for a slight overlapping of the phases required to illustrate a complete stride or movement.

The time-intervals of exposures varied from the one-hundredth of a second to several seconds. A record of these time-intervals was kept by the chronograph —a well-known instrument, used in every physiological laboratory—it comprises a revolving cylinder of smoke-blackened paper, on which, by means of successive electric contacts, a style is made to record the vibrations of a tuning-fork, while a second style marks the commencement of each successive exposure. The number of vibrations occurring between any two exposures marks the time.

The tuning-fork made one hundred vibrations in a second of time. To ensure greater minuteness and accuracy in the record, the vibrations were divided into tenths, and the intervals calculated in thousandths of a second.

For the purpose of determining the synchronous action of the electro-exposers while making a double series of exposures, the accuracy of the time-intervals as recorded by the chronograph, and the duration of the shortest exposures used in the investigation, the two cameras of twelve lenses each were placed side by side, and the exposers were connected through their respective magnets with the motor-clock by separate lengths of one hundred feet of cable. The lenses of the two cameras were pointed to a rapidly revolving disc of five feet diameter. The surface of the disc was black, with thin white threads radiating from the centre to the edge.

A microscopic examination of the two series of resulting negatives failed to prove any variation from the synchronous action of ten of the duplicated series, and in the two others there was a discrepancy in the simultaneity by a few ten-thousandths of a second; a result sufficiently near to synchrony for any ordinary use.

The shortest exposures made at the university were in about the one six-thousandth part of a second, and in this time details in black and white drapery were obtained on the same negative. Such brief exposures were in this class of work rarely needed. Some horses, galloping at full speed, will cover nineteen yards of ground in a second of time, or a full mile in a hundred seconds or less. At this speed a foot, recovering from its rest, will be thrust forward with an occasional velocity of more than forty yards in a second. During the one-thousandth part of a second, the body of the horse may move forward about seven-tenths of an inch, and a moving foot perhaps one and a half inches—no very serious matter for ordinary requirements.

A knowledge of the duration of the exposure was in this investigation of no value, the aim always being to give as long a time as the rapidity of the action would permit, with a due regard to essential sharpness of outline and distinctness of detail.

Although the one six-thousandth part of a second was the most rapid exposure made on this occasion, it is by no means the limit of rapidity in mechanically effected photographic negative exposing; nor does the one-hundredth part of a second approach the limit of time-intervals. Marey, in his physiological experiments, has recently made successive exposures with far less intervals of time; and the author has devised, and hopes some day to make use of, an apparatus which will photograph twenty consecutive phases of the vibration of an insect's wing, even assuming as correct a quotation by Pettigrew from *Nicholson's Journal*, that a common house-fly will make, during flight, seven hundred vibrations in a second—a number probably much in excess of the reality.

It may be here appropriately mentioned that Marey, in 1882, discarded his original "graphic" method of analyzing motion for the more effective photographic process.

At the Palo Alto investigation a series of negatives was frequently made by threads stretched across the track of the animal. The thrust against each of these threads in succession completed an electric circuit, and effected a photographic exposure; the thread was subsequently broken by the progress of the animal. Some seriates were made by the wheel of a vehicle, to which a horse was attached, depressing wires; each depression completed a circuit and effected the exposure of a negative. For small animals and birds, and for movements without regular progressive motion, the motor-clock was necessarily used.

The rapidity of the transmission of nervous sensation was experimented with. The explosion of a small torpedo in close proximity to an animal or bird, started the motor-clock, and commenced a series of exposures. An example of its effect may be seen in Plate 781 of "Animal Locomotion."

PRELUDE TO ANALYSES.

When an animal is carrying itself forward by any system of regular motion, its limbs, in their relation to the body, have alternately a progressive and a retrogressive action; their various portions are accelerated in comparative speed as they extend downwards to the feet, which are subjected to successive changes from a total cessation of movement to a varyingly increased velocity in comparison with that of the body.

Photographic analysis has demonstrated that quadrupeds employ, on the surface of the ground, eight different regular systems of progressive motion. They are—

1. The walk.
2. The amble.
3. The trot.
4. The rack.
5. The canter.
6. The transverse-gallop.
7. The rotatory-gallop.
8. The ricochet.

In this enumeration crawling is omitted, it being simply a modified system of walking, and subject to the same rules.

Leaping or jumping by the use of all four of an animal's legs can be regarded only as an accidental interruption to regular progress.

All other methods which may be occasionally employed by, or which it is possible for an animal to use in terrestrial locomotion, may be considered as abnormal movements.

The differences between the step and the stride of an animal are not always clearly understood.

A "step" is an act of progressive motion, in which one of the supporting members of the body is lifted from the ground, thrust in the direction of the movement, placed again on the ground, and caused to reassume, either wholly or in part, its proper functions of supporting and propelling the body.

A "stride" is a combination of actions in progressive motion, which requires each one of the supporting members of the body, in the exercise of its individual functions, to be—either alone or in association with another supporting member—lifted from the ground in its regular sequence, thrust in the direction of the movement, placed again on the ground, and caused to reassume the same relative position to the body and to the other limbs as it occupied at the commencement of the notation.

The normal stride of a biped consists of two uniformly executed steps.

Shakespeare recognizes this fact in *The Merchant of Venice*, act iii, sc. 4—

"I'll . . . turn two mincing steps
Into a manly stride"

The normal stride of a quadruped, while using four limbs as supports, during locomotion, consists of four steps.

These steps may occur singly, and at approximately regular periods of time, as in the walk; singly, and at irregular periods, as in the amble, the canter, or the gallop; or in pairs, as in the trot or in the rack.

To facilitate a study of the various systems of support and propulsion employed by an animal during the execution of any of its regular gaits, symbols have been adopted to designate the feet which during the instant of a particular phase are actually engaged in one or both of these special functions

These symbols are for—

	Left.	Right.
Anterior, or fore feet	△	▲
Posterior, or hind feet	○ ...	●

Denotes that the left fore-foot and the right hind-foot are at that instant being used to support or to propel the body.

Denotes a transit without, at that instant, the actual support of any one of the feet.

In the diagrams, arrow-heads indicate the direction of the movement. The sequences of the phases are regularly numbered.

The intervals of time or of distance between any two phases of the diagrams are not there recorded, nor is the precise locality of any foot indicated. These facts can only be ascertained by reference to the illustrations from which the diagrams are constructed.

In the execution of the eight distinct systems of regular progressive motion, animals employ fifteen different methods of temporary support, all of which are, under various conditions, made use of by the horse. They are—

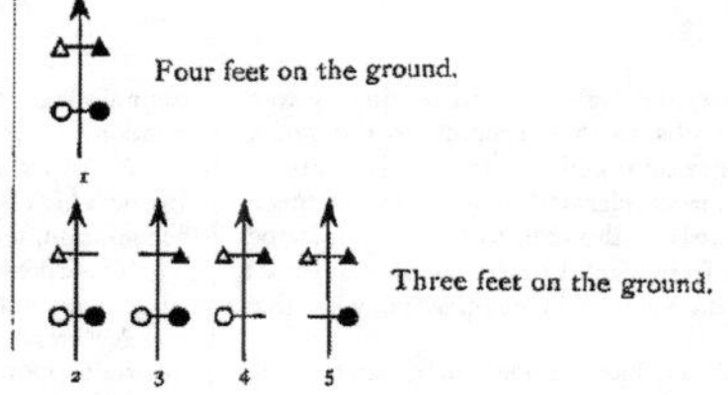

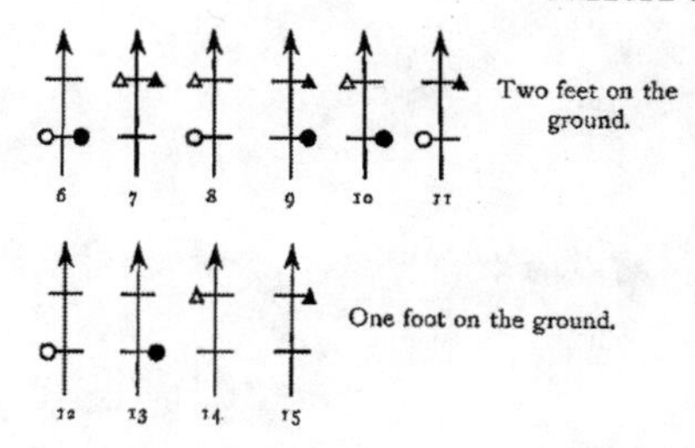

In the illustrations the sequences of phases are arranged in the direction of the movement, or are marked by an arrow.

When the consecutive phases of a movement have been synchronously photographed from two or more points of view, the fore-shortened phases are arranged in the same direction as the laterals; the corresponding phases of any series—if the small size of some of their numbers prevent their being readily seen—can be ascertained by counting the succession.

The time-intervals between the phases are marked in thousandths of a second; the time of a complete stride, approximately, in hundredths of a second.

The distance measurements are, approximately, given in inches, and in metres roughly calculated on the basis of forty inches to a metre.

In the analyses of the gaits the word "spring" must not always be taken to imply a leap; it is frequently used as a convenient term to indicate the last impulse of a foot prior to its being lifted from the ground.

During very rapid motion by a good horse, the aggregate of the body preserves a nearly horizontal line, the period of transit without support being usually too brief for the attraction of gravitation to have much effect, and the body of the animal is nearer to the ground by the height of the fetlock or pastern-joint than when standing at rest.

THE WALK.

Of the various methods of animal motion, the walk claims our first consideration; it is characterized by an immutable sequence of limb movements, common alike to man and beast, and there is little doubt of its having been the primitive system of locomotion employed, on their evolution, by all the terrestrial vertebrates.

The law governing this method of progress is, that the naturally superior or stronger limb takes precedence of its inferior lateral limb in being lifted, thrust forward, and again placed on the ground.

During the walk of a quadruped whose constant habit is to travel on the surface of the ground, and to employ all four of its feet for the purposes of support and propulsion, the successive foot-impacts, assuming the notation to commence with the landing of ○, will be—

▲ being, of course, followed by ○ in the next stride.

When a horse is standing with the weight of the body equitably distributed over his four legs, and under these conditions commences to walk, the initiatory movement will invariably be made with a hind-foot; the lateral fore-foot will next follow, and under the normal conditions of regular progress this fore-foot will be lifted in advance of the suspended hind-foot being placed on the ground.

The rapidity with which any one foot follows any other foot, or the duration of its contact with the ground, vary greatly, not only with different species of animals, but also with the same animal under apparently similar conditions.

Series 1 illustrates twelve consecutive phases of two steps, or one half of a complete stride of a horse, walking at a speed of about four and a quarter miles an hour. In phase 1, although ○ is still flat on the ground, it has practically relinquished its function of support, and, as in 2, 3, and 4, that duty is imposed on the diagonals ● △; 4 exhibits ▲ a fraction of an inch only above the ground, but it assists the labours of ● △ in 5.

In 6 △ has broken the alliance, in this phase, and also in 7, 8, and 9, the right laterals alone furnish the needed support. ○ comes to the aid of these laterals in 10 just as △ is being advanced beyond ▲. The toe of ●

still lingers on the ground in 12; a phase fractionally in comparative advance of O in 1.

One half of the stride is now completed.

Assuming that no interruption takes place, and the horse to be walking, under the usual conditions, on level ground, the remaining half—with a substitution of the right feet for the left—will be executed in practically the same manner.

This analysis determines the successive methods of support afforded by the feet of a horse during a normal stride of the walk to be—

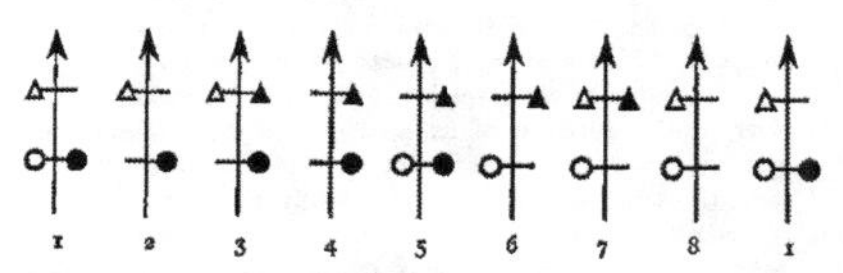

The notation may, of course, be commenced at any phase, but in the normal walk of a horse it will invariably be found that the support is thrown, during one stride, twice on the laterals, twice on the diagonals, twice on two fore feet and one hind-foot, and twice on two hind feet and one fore-foot; eight different systems of support.

Series 2 illustrates twelve consecutive phases of a powerful draught horse pulling a dead weight of perhaps one thousand pounds, requiring a continuous strain.

The synchronous fore-shortenings are on p. 29.

Series 3 demonstrates the walk of a thorough-bred Kentucky mare, who is also represented in the canter and the gallop. On p. 41 are five phases each, of a fast walk, and a slow trot; the last phase of the walk indicates a tendency to increase the pace to an amble. The plate may be found useful in demonstrating the transitions to and from the central phases of each line, they having some points of resemblance.

When an animal is walking very slowly, the supports are not furnished alternately by two and by three feet, as in the normal walk, but by alternations of three and of four feet, each foot is placed in regular succession on the ground in advance of its preceding foot being lifted therefrom. Had the ass in series 5 been walking a little more slowly, four feet on the ground would have been seen in phases 1, 7, and 12. An animal grazing in the fields affords an illustration of a very slow walk, and a good opportunity of studying the sequence of foot-fallings.

The ox, goat, and hog, as representatives of double-toed, or cloven-footed animals; and the elephant, Bactrian camel, lion, dog, raccoon, and capybara as representatives of soft-footed quadrupeds, will be found, in their respective seriates, to follow, while walking, the same sequence of foot-fallings as that disclosed by the horse.

A noteworthy confirmation of the law governing the walk was found in the case of a child suffering from infantile paralysis, whose only method of locomotion was by the use of her limbs exactly in the manner of a quadruped. In her progress it was revealed that not only was the regular system of limb movements used, but the support of the body devolved, in their proper sequence, on the laterals and on the diagonals. The diagram in this chapter is as faithful a representation of the consecutive impacts by that child as it is of those by an elephant, a turtle, or a mouse.

In our enumeration of movements, crawling is classified as a method of walking. Series 17 illustrates a strong, healthy child crawling on her hands and knees. Phase 4 clearly demonstrates support on the diagonals alone; that afforded by the laterals is not so easily recognized; the succession, however, is indisputable

With man walking erect, as we find in seriates 18 and 19, the culmination of the swing of the arm must be considered as the equivalent of placing its hand on the ground. It will be seen in 1, series 18, that although the right foot is not yet flat on the ground, and the toe of the left foot remains in contact therewith, the right arm has commenced its forward thrust, which terminates in 6, before the heel of the left foot reaches the ground. In 7, the right hand is on its backward swing, while the left has commenced its forward motion. Attempts were made to obtain visible evidence of the tendency of a bird's wing while using its legs in walking; the resulting information was inconclusive. In series 20 we have an animal that apparently disregards the law governing the walk. Although the ape family, during their progress on the surface of the ground, are accustomed to use all four of their limbs as supports, their constant habit of climbing has so developed the strength of their anterior limbs, or arms, that they have become the superior, and consequently, in their movements, usually take precedence of their laterals.

If, while a horse is walking, two moving feet are seen respectively in advance of, and to the rear of the supporting legs, they are diagonals; if two moving feet are seen under the body, between the supporting legs, they are diagonals, as disclosed by phases 2 and 7, series 1. In series 20, phases 1 and 3, this rule is reversed; but it is not invariably followed. An ape will occasionally walk on all-fours, with the same order of foot-fallings as that which characterizes the rotatory-gallop.

The family of apes, when climbing, make prior use of the stronger lateral, as may be seen in series 21, representing a baboon climbing a pole.

The sloth would find its horizontally suspended walk difficult to execute with any relaxation of diagonal support, as series 22 demonstrates.

The movements of animals in their relation to design in Art requires far broader treatment than is possible in the present volume; its province in this important matter will, therefore, be confined to a superficial review of the expression given to some of the movements, as illustrated by a few examples of ancient and modern times. It is worthy of note that the presumed most ancient relic yet discovered of artistic design represents the quadrupedal walk scientifically correct. The position of the limbs of the reindeer, in the well-known etching by some prehistoric artist, is precisely the same as photographed from nature in series 1, phase 8.

The inflexible laws of an all-powerful priesthood, and the superstitions of a docile people, prohibited the Egyptian artist from giving more than one expression to the walk of a quadruped, except to that of the horse; the exception is probably due to the fact of the horse being an unknown animal in Egypt when the decree was made. The phase adopted can readily be seen by watching a cow grazing until it reaches a stage of progress when, with all four feet on the ground, the right legs incline forward from their base and the left legs incline backward, the direction usually being from left to right. This phase of the walk was used by the Egyptians for asses, oxen, jackals, porcupines, and other animals, in endless repetition, in their manuscripts and decorative paintings, and in the carvings on their temples and sarcophagi The common Egyptian interpretation of the walk is represented by the photograph of the ass, p. 83.

Although in Egyptian art the horse is far less skilfully drawn than are other animals, the expression given to his walk is correct; the phase usually adopted resembles that of 2, series 1. Whether the Assyrians derived their art inspirations from the Egyptians, the Egyptians from the Chaldæans, or whether they were all originally taught by a race of whom we have no remains or tradition, will probably never be determined. It is evident that there was much communication between the people of the second Assyrian empire and the Egyptians. There are strong points of resemblance in their interpretations of animal movements; the bent knee in the walk of the former, however, is not usually found on the Nile, except in illustrations of the horse.

In Hamilton's "Early Greek Vases" appears a design of Diomedes and Ulysses, presenting to Nestor the horses of Rhesus, the horses are apparently copied from some Egyptian design.

In the perfection of their work, some of the later Greeks were inclined to represent the walk more, perhaps, as they thought it ought to be than as it really is.

On the arch of Titus; the column of Trajan, and in many of their statues, the Romans seem to have been indifferent to their interpretation of this action.

A certain phase of the trot has been very generally used by painters and sculptors of the horse to represent the action of walking. It is frequently difficult, both in ancient and modern art, to determine whether it is the intention of the designer to indicate a trot of ten miles an hour or a walk of one-third of that speed.

The statue of Marcus Aurelius, at Rome, is a remarkable instance of the failure of a sculptor to express his obvious intention. The pose of the emperor, and other circumstances, point to a deliberate motion of the horse, which is not confirmed by its method of progress.

Many of the equestrian statues of Europe and America are, virtually, reproductions of Marcus Aurelius, and represent the legs of the animal performing a lively trot of eight or ten miles an hour, while the rider sits with as calm a repose as if taking part in a solemn procession. The larger figure, p. 41, very closely reproduces the phase of motion, selected by the sculptor, for the horse on which the Roman emperor is seated.

This apparent indifference, or lack of discrimination by the artist, was shown in the reliefs on the column of Theodosius, erected at Constantinople in the fourth century. Two heavily-laden pack-horses in a procession, one immediately in front of the other, are represented, the one trotting, the other walking; many other animals, oxen, camels, elephants, etc., are intended to be represented walking, to some of which the artist gave a correct interpretation, to others, an erroneous one; the greater number, however, were strictly correct.

The bronze horses over the portals of St. Mark's at Venice, are fine examples of a careful study of natural action.

Of the great masters of the fifteenth and two succeeding centuries, Donatello and Verrocchio are the most pronounced in their complete understanding of this movement, as their respective statues at Padua and Venice afford ample proof. Albert Durer, in "The Knight, Death and the Devil," leaves a singular memento of his carelessness in giving effect to his avowed intention. One of the greatest Austrian artists of this century, in companion pictures, each of a procession in which women and children are taking part, has the central figure of one picture on a horse walking, of the other, on one trotting.

A celebrated animal painter of France, in a picture so meritorious as to be considered worthy of a place in the national collection, depicts several oxen yoked to a plough; from the vigorous efforts of the driver to goad them on, they are supposed to be making very slow progress, but one, only, of the animals is walking; the others, with probably the same inclination, are moving with a variety of gaits.

Error in the interpretation of the quadrupedal walk had become so predominant, that when Meissonier exhibited his picture of "1814," he was much ridiculed by the artists and critics of Paris for having—as they supposed—misrepresented that action. In 1881 the great painter assembled his colleagues of the Academy in his studio for the purpose of convincing them, as he himself announced on the occasion, that "the Sun had now been invoked to prove the truth of Meissonier's impression." It is unnecessary to point out the phase selected by the artist for the leading horse of his picture.

The "Roll Call" affords another well-known example —to the astonishment of the critics at the time—of a careful study of the walk.

The walk of a quadruped being of a slow unexciting character, is perhaps the reason why few references are made to it, by name, in poetry, or even in general prose compositions.

Shakespeare uses the word, metaphorically, in *Hamlet*, i. 1—

> "The morn in russet mantle clad,
> Walks o'er the dew of yon high eastern hill."

Milton, in "Paradise Lost," vii., says—

> "Among the trees, in pairs they rose and walked."

And Swift, in a voyage to the Houyhnhnms, reports that Gulliver "saw a horse walking softly in the field."

Authors, as a rule, indicate the pace by some inferential word. Thus Wordsworth, in "The Old Cumberland

Beggar": "the sauntering horseman." Longfellow, in "Tales of a Wayside Inn"—

"A jaded horse, his head down bent,
Passed slowly, limping as he went."

Dickens, "David Copperfield," iii.: "The carrier's horse was the laziest horse in the world . . . and shuffled along with his head down." Morris, "Bellerophon at Argos"—

"The slow tramp of a great horse soon they heard."

Barham, in "The Lay of Aloys," alludes to the walk of a cat under difficulties—

"From his lurking place,
With stealthy pace,
Through the long-drawn aisle he begins to crawl,
As you see a cat walk on the top of a wall
When it's stuck full of glass, and she thinks she shall fall."

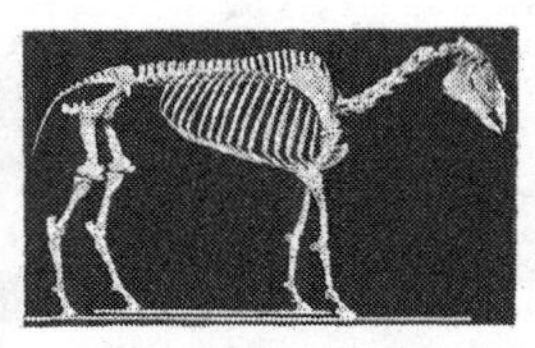

A PHASE IN THE SLOW WALK OF A HORSE.

THE WALK.

Series 1.

A HALF-STRIDE, PHOTOGRAPHED SYNCHRONOUSLY FROM TWO POINTS OF VIEW.

Horse "Eagle."

Length of complete stride, 88 inches (2·20 metres). Time-intervals ·052 second.
Approximate time of complete stride, 1·20 seconds. Strides to a mile 720. Speed per mile 14 minutes.

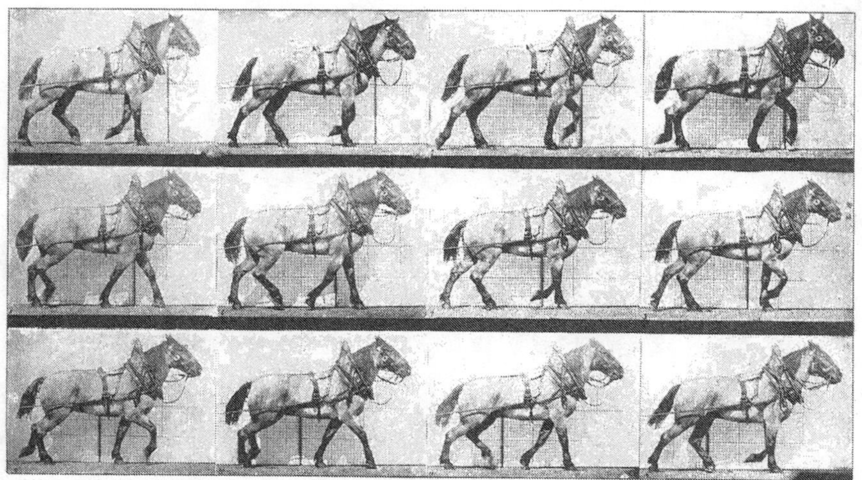

SERIES 2

ONE STRIDE DRAGGING A HEAVY DEAD-WEIGHT

Horse "Billy"

Length of stride 60 inches (1 50 metres). Time-intervals 111 second. Approximate time of complete stride 1 19 seconds

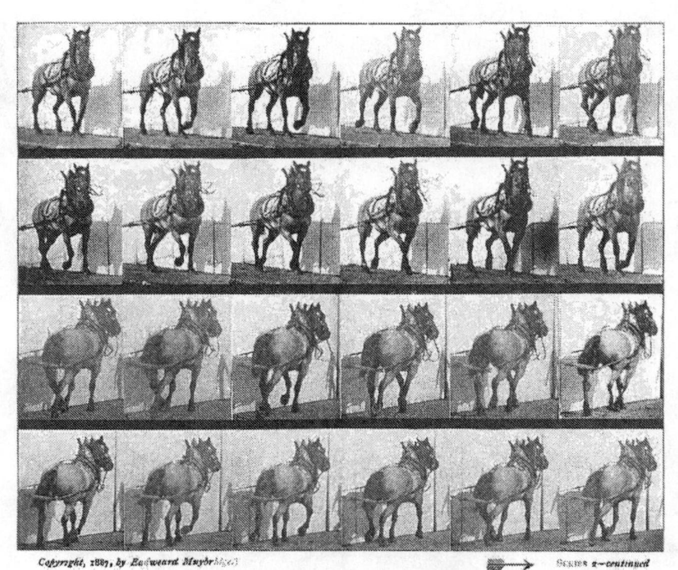

 SERIES 2—*continued*

ONE STRIDE: DRAGGING A HEAVY DEAD-WEIGHT.

Horse "Billy"

Synchronous foreshortenings of corresponding phases in series 2 arranged in the same consecutive order.

THE WALK.

SERIES 3

A HALF-STRIDE, PHOTOGRAPHED SYNCHRONOUSLY FROM TWO POINTS OF VIEW

Thorough-bred Mare "Annie"

Length of complete stride 84 inches (2·10 metres) Time-intervals ·044 second
Approximate time of complete stride ·95 second

THE WALK.

Series 4.

ONE STRIDE IN TEN PHASES, PHOTOGRAPHED SYNCHRONOUSLY FROM THREE POINTS OF VIEW

Horse "Elberon."

Length of stride. 86 inches (2·15 metres). Time-intervals ·126 second Approximate time of stride 1·17 seconds

SOME CONSECUTIVE PHASES OF THE WALK FROM SERIES 2.

Horse "Billy"

THE WALK.

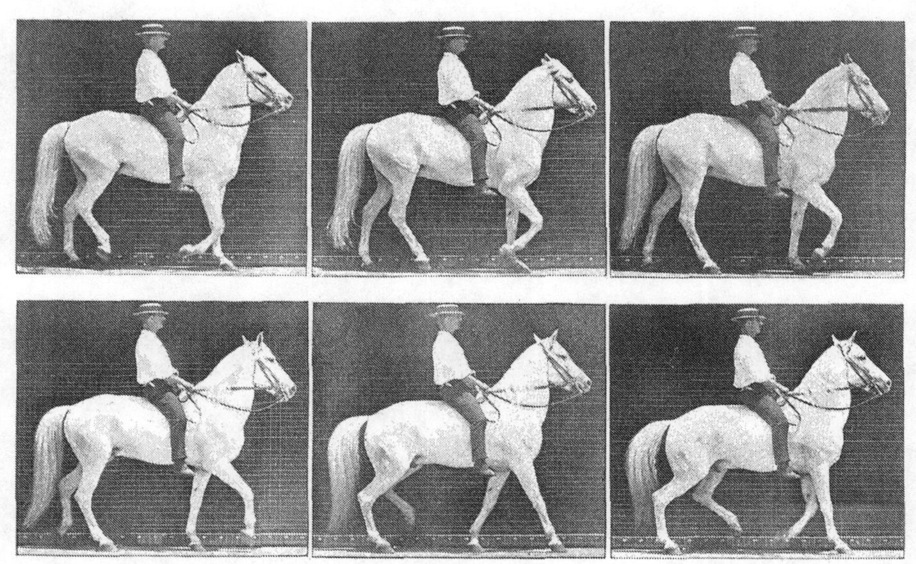

SOME CONSECUTIVE PHASES OF THE WALK

Horse "Clinton"

THE WALK.

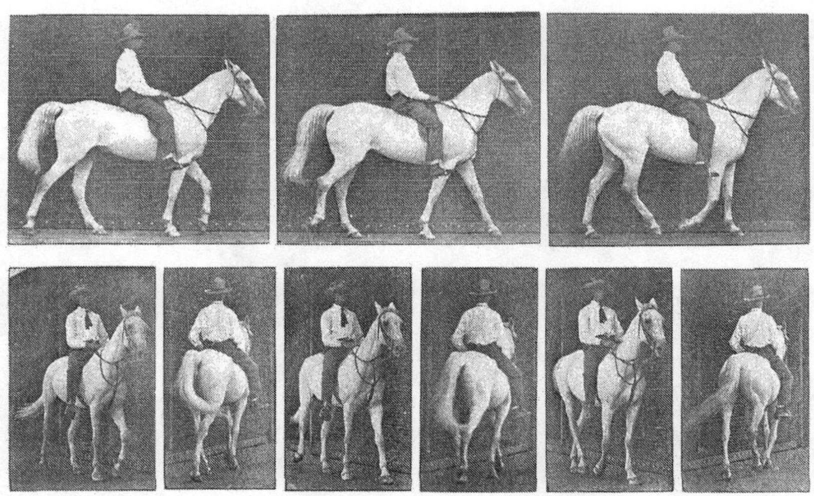

THREE PHASES OF THE WALK FROM SERIES 4
EACH PHASE PHOTOGRAPHED SYNCHRONOUSLY FROM THREE POINTS OF VIEW
Horse "Elberon"

THE WALK.

SOME CONSECUTIVE PHASES OF A FAST WALK.

COMPARED WITH SOME CONSECUTIVE PHASES OF A SLOW TROT.
The right feet are indicated by dots near the pasterns.

A PHASE OF THE TROT

THE WALK.

Copyright, 1887, by Eadweard Muybridge]

SERIES 5

A HALF-STRIDE IN SEVEN PHASES.

The Ass

Length of stride 44 inches (1 10 metres). Time-intervals 094 second. Approximate time of stride 1 06 seconds.

THE WALK

A HALF-STRIDE IN THIRTEEN PHASES

The Ox

Length of complete stride: 84 inches (2 10 metres). Time-intervals 048 second Approximate time of complete stride 1 15 seconds

THE WALK

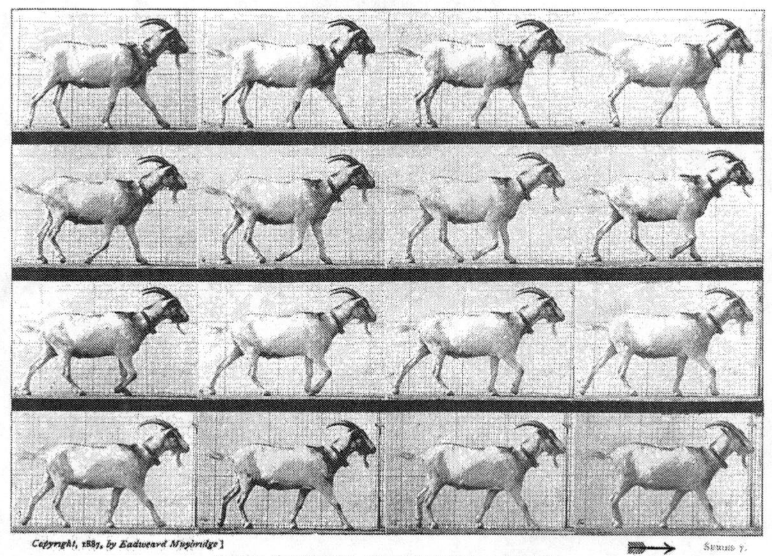

Series 7.

A HALF-STRIDE IN THIRTEEN PHASES.

The Goat.

Length of complete stride 48 inches (1 20 metres). Time-intervals · 043 second Approximate time of complete stride 1 03 seconds.

THE WALK.

SERIES 8

ONE STRIDE.

The Hog

Length of stride, 44 inches (1 10 metres). Time-intervals: 077 second. Approximate time of stride: 85 second

 SERIES 9

IRREGULAR WALKING IN CONFINEMENT.

The Tiger

Time-intervals 119 second

IRREGULAR: WALKING IN CONFINEMENT.

The Lion.

Time-intervals ·076 second

THE WALK.

Series 11

BREAKING INTO A GALLOP

The Cat

Time-intervals 031 second. Approximate time of complete series 71 second.

 SERIES 12

ONE STRIDE, NOT QUITE COMPLETED

The Elephant.

Height of animal: 100 inches (2.50 metres) Length of stride: 112 inches (2.80 metres)

Time-intervals: .063 second. Approximate time of stride: 1.32 seconds.

THE WALK.

SERIES 23.

A HALF-STRIDE IN FOURTEEN PHASES.

The Bactrian Camel.

Length of stride 100 inches (2 50 metres). Time-intervals 057 second Approximate time of complete stride 1·48 seconds.

THE WALK.

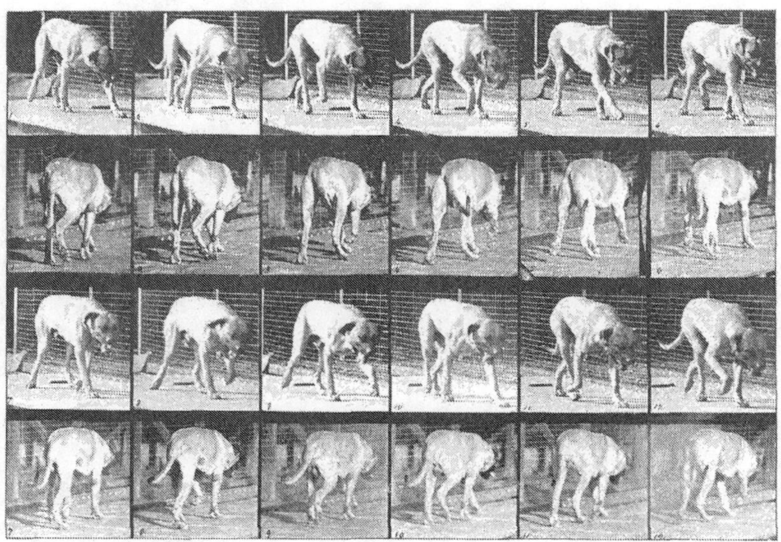

Copyright, 1887, by Eadweard Muybridge.] Series 14

ONE STRIDE IN TEN PHASES, PHOTOGRAPHED SYNCHRONOUSLY FROM TWO POINTS OF VIEW

The Dog (Mastiff)

THE WALK.

SOME PHASES IN THE WALK OF A DOG FROM SERIES 14.

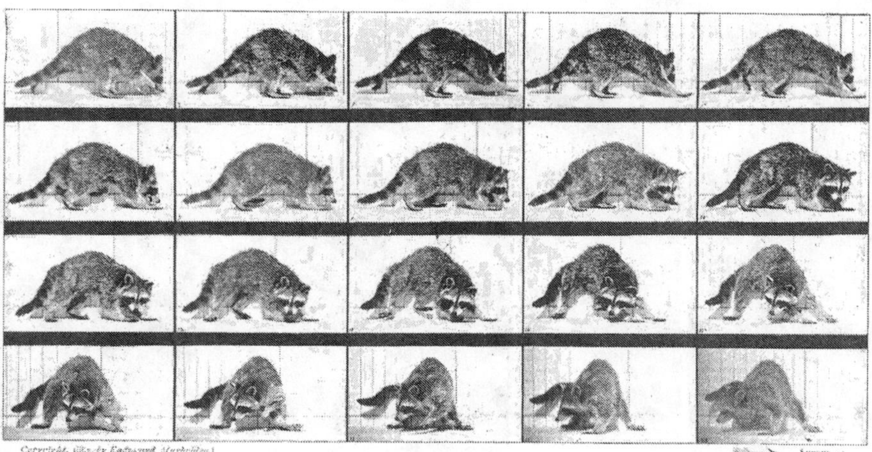

SERIES 15.

TURNING AROUND.

The Raccoon

Time-intervals. ·032 second.

K

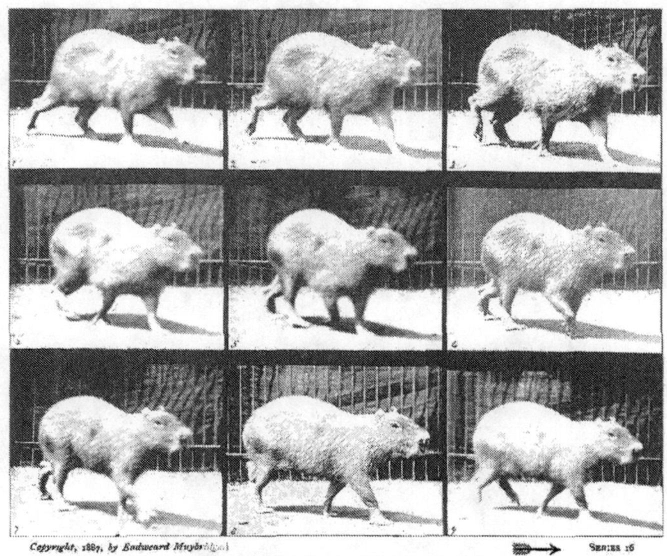

 SERIES 16

A HALF-STRIDE IN SEVEN PHASES.

The Capybara.

SERIES 17.

CRAWLING ON HANDS AND KNEES: ONE STRIDE IN SIX PHASES.

The Child

DEMONSTRATING THE LAW GOVERNING THE CONSECUTIVE ACTION OF THE LIMBS IN THE PRIMITIVE METHOD OF TERRESTRIAL PROGRESSIVE MOTION BY VERTEBRATES.

Time-intervals. ·169 second.

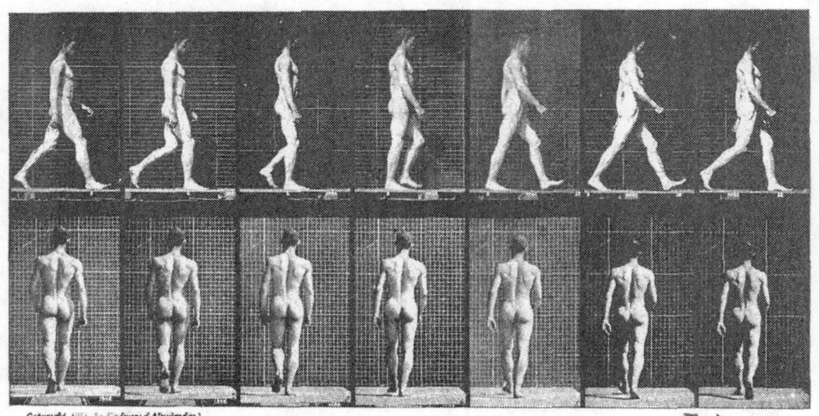

Series 18

A HALF-STRIDE, PHOTOGRAPHED SYNCHRONOUSLY FROM TWO POINTS OF VIEW.

Man (Athlete).

Demonstrating the Law governing the Consecutive Action of the Limbs in the Primitive Method of Terrestrial Progressive Motion by Vertebrates

Length of complete stride · 72 inches (1·80 metre) Time-intervals ·083 second Approximate time of complete stride. ·95 second

 Series 19

A HALF-STRIDE, PHOTOGRAPHED SYNCHRONOUSLY FROM TWO POINTS OF VIEW.

ENLARGED FROM A PHOTO-ENGRAVING.

Man (Athlete).

Time-intervals .069 second

Copyright, 1887, by Eadweard Muybridge.

SERIES 20

ONE STRIDE IN FOURTEEN PHASES

DEMONSTRATING THE LAW GOVERNING THE CONSECUTIVE ACTION OF THE LIMBS IN THE PRIMITIVE METHOD OF VERTEBRAL PROGRESSIVE MOTION, THE ANTERIOR LIMB, OR ARM, BEING THE SUPERIOR

The Baboon

Length of stride: 41 inches (1.02 metres). Time-intervals: .045 second. Approximate time of stride: .61 second

SERIES 92

A HALF-STRIDE IN NINE PHASES.

CLIMBING, OR WALKING IN A VERTICAL DIRECTION BY SUSPENSION.

The Baboon

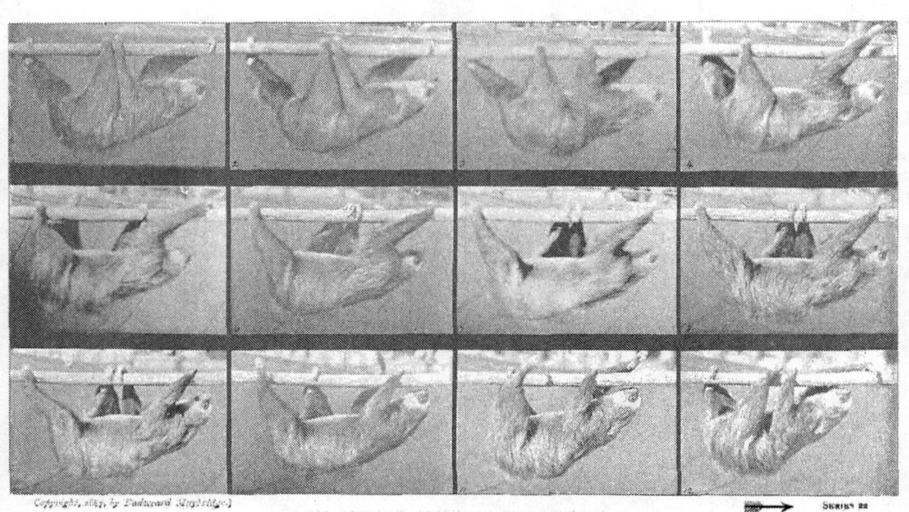

Series 22

ONE STRIDE, NEARLY COMPLETED

Walking Horizontally by Suspension

The Sloth

Copyright, 1887, by Eadweard Muybridge] Series 93

A HALF-STRIDE IN NINE PHASES

The Adjutant

THE WALK.

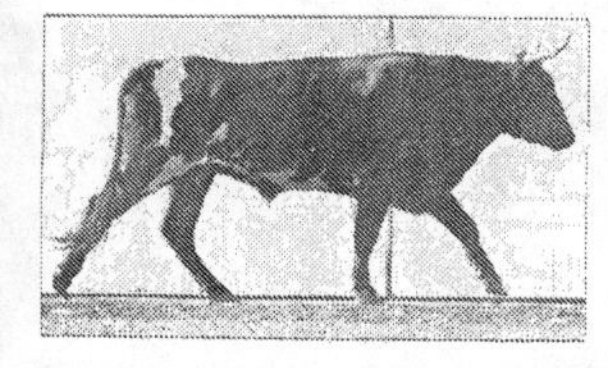

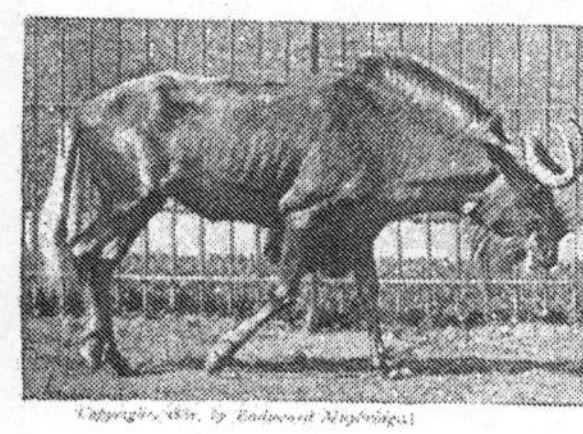

(Copyright, [illegible], by Eadweard Muybridge.)

1 2 3 4

SOME PHASES OF THE WALK.

AS EXECUTED BY THE ASS, THE OX, THE HOG, THE GNU, AND THE BUFFALO OR BISON

SOME CONSECUTIVE PHASES OF THE WALK OF A HORSE, PHOTOGRAPHED FROM THE REAR

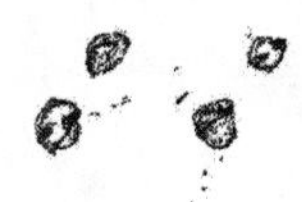

Copyright, 1887, by Eadweard Muybridge.

SOME PHASES IN THE WALK OF THE LIONESS, THE LION, THE TIGER, THE DOMESTIC CAT, THE CAMEL, COMPARED WITH THE CRAWLING OF A CHILD, THE SUSPENDED WALK OF THE SLOTH, AND THE WALK OF THE BABOON REPRESENTING THE APE FAMILY.

THE WALK.

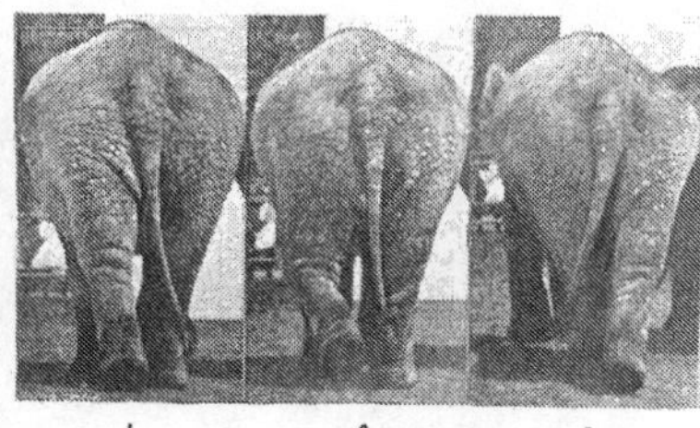

SOME PHASES IN THE WALK OF THE TIGER, THE JAGUAR, AND THE LION, ALL IN THE ACT OF TURNING ROUND. THE VERTICALLY SUSPENDED WALK OF THE BABOON. THE WALK OF THE CAPYBARA, REPRESENTING THE RODENTS. AND THREE CONSECUTIVE PHASES OF THE WALK OF AN ELEPHANT, PHOTOGRAPHED FROM THE REAR

THE AMBLE.

THE amble is a development of the walk into a mode of progress from which a higher rate of speed may be obtained. Practically, it is an accelerated walk; it has the same sequence of foot-impacts, but from their more rapid succession, a hind-foot and a fore-foot are alternately lifted from the ground in advance of its following foot being placed thereon.

This procedure results in throwing the duty of support alternately on one foot and on two feet. A hind-foot and a fore-foot successively furnish the single support; diagonals and laterals alternate in supplying the duplex support.

Series 24 demonstrates how this movement is consummated.

In 1 the support devolves on ● ▲, with—as in the walk—○ △ suspended between them. In 3 ● is lifted in advance of ○ being landed, which is, however, on the ground in 4, where ○ ▲ jointly sustain the weight of the body; the bent knee of ▲ indicates that ○ will soon have to perform its labours alone, as it is doing in 5; △ soon comes to its assistance, and in 6 the left laterals assume the responsibility which in 1 devolved on the right laterals. One-half of the stride is now completed, and so far all has gone as it should; had the remaining moiety been executed with similar precision, there would have been no fault to find. In Pennsylvania, ambling horses are not so abundant as they are in Kentucky, California, and some other countries; the only horse capable of ambling, and obtainable, was the one here represented, who neglected to use his legs in the orthodox manner during the second half of the stride.

The six consecutive phases used as an illustration of this gait may, however, be accepted as perfectly characteristic of the complete movement, which may be recorded in the diagram as—

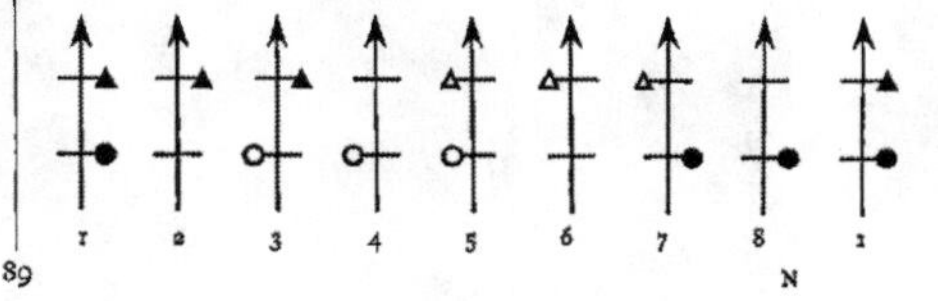

This motion is perhaps better scientifically demonstrated in series 26, which represents a complete stride by a first-class ambling horse, photographed at Palo Alto during the summer of 1879 The horse not having been of a suitable colour for the background, the outlines were carefully filled in to give the figure more distinctness, and a dot added to distinguish the right feet from the left. The stride is somewhat more than completed in phase 11. No record of the speed was taken, but it probably was about seven miles an hour.

Series 27 illustrates twenty-four phases of one nearly completed stride of an elephant while progressing at as fast a speed as vigorous persuasion could induce—equivalent to a mile in somewhat less than seven minutes.

The gait resorted to was the amble. In phase 10 the weight of the body devolves on ▲; 12 demonstrates the assistance rendered by ○; the bend of the knee in 14, which is more pronounced in 15, determines ○ to be practically furnishing exclusive support for a brief period, which function is shared by △ during several following phases. In 21 △ assumes the entire responsibility until 23, when the animal is again fairly on the diagonals.

The diagram of the stride of a horse is equally applicable to one by the elephant.

The walk and the amble are probably the only two gaits used by the elephant in his natural state. Oriental paintings and carvings may not be very trustworthy sources of information, but so far as they have been examined by the author, they corroborate this supposition.

It is very remarkable that, although the amble is the most comfortable to the rider, of all the gaits which are natural to the horse, or to which he has been trained, it is now, in Great Britain, either entirely unknown, or has lapsed into disfavour. It is perhaps more remarkable that many writers on the horse and horsemanship should have confused this delightful, easy motion, with that disagreeable jolting gait, appropriately termed the rack, or, as it is ambiguously called by some horsemen, "the pace."

It would seem plausible that the very earliest riders of the horse would very soon discover the steady and comparatively rapid motion of the amble, just as the North American Indians have, whose acquaintance with the animal does not date back much more than two centuries.

The gait was evidently well known to the ancients. On the walls of Karnak, the great Rameses is represented on his return from the wars with prisoners; he is standing in a chariot drawn by two ambling horses. The phase corresponds with one occurring between 4 and 5, series 23.

Horace, in his "Epistles," as translated by Francis, alluding to a retired citizen who enjoyed comfort, says—

> "On horse-back now he ambles at his ease."

Vegetius, in the fourth century, writes of the "ambulatura" being the favourite gait of the wealthy and indolent Romans, and of the care they bestowed on their horses to make them perfect in it.

Illuminated manuscripts of the tenth and later centuries —if they may be considered as reliable evidences—prove that ambling was constantly practised by the Anglo-Saxons and Normans, especially in the diversions of hunting and hawking. In the Bayeux Tapestry "one" arrives before Duke William on an ambling horse.

We have the testimony of Chaucer that the Canterbury Pilgrims made their journey on amblers. The illuminated manuscript of the "Canterbury Tales," in the Ellesmere Collection, confirms the poet's assertions. The "good wyf" is seated, masculine fashion, on an "amblere" in a phase exactly corresponding with 2, series 24, the horses of many others of the company are also represented as practising the same motion, the favourite position more or less resembling that of 3, with △ or ▲ alone on the ground. The prologue has—

> "A good Wyf was ther bisyde Bathe . . .
> Vppon an amblere esely sche sat;"

and "The Tale of Sir Thopas," Fyt I.—

> "His steed was al dappul gray
> Hit goth an ambel in the way,
> Ful softely and rounde."

Gower, in "Confessio Amantis," has—

> "On fayre ambulende hors thet set,"

and—

> "Thei set him on an ambuling palfray."

In the "Morte d'Arthur," translated from the French in the fifteenth century by Sir Thomas Malory, the "softe ambuler" is often alluded to.

In "The regulations and Establishment of the Household of Algernon Percy the Fifth Earl of Northumberland. Begun anno 1512" there occurs—

> "Item, palfreys for my ladys, to wit, one for my lady, and two for her gentill-women.
> "An amblynge horse for his lordship to journey on dayly.
> "A proper amblyng little nagg for his lordship when he gaeth on hunting or hawking"

Polydore Virgil, in the fifteenth-sixteenth century, says English horses "are not given to the trot, but excel in the softer paces of the amble."

Holinshead, sixteenth century, says, "The Irish hobbie is easie in ambling, and verie swift in running."

Shakespeare was evidently perfectly familiar with this pleasant mode of progress; he uses it as a metaphor in "As You Like It," iii. 2, and in "Much Ado about Nothing," v. 1.

Ben Jonson, also, in "Every Man in his Humor," metaphorically says, "Out of the old hackney pace, to a fine easy amble."

Cervantes, in "Don Quixote," ii. 40, as translated by Skelton, gives an admirable description of this motion: "This horse . . . ambles in the ayre, without wings, and he that rides upon him may carry a cup full of water in his hand, without spilling a jot; he goes so soft and so easie."

Gervase Markham, a celebrated authority on horses, writing in 1615, says, "The ambler . . . is the horse of the old man, the rich man, and the weak man."

Gibbon, "Roman Empire," lviii., speaks of the war-horse of the knight, until approaching danger, being usually led by an attendant; the knight himself "quietly rode a pad or palfrey of a more easy pace."

Cowper, in "Retirement," says—

> "To cross his ambling pony day by day,
> Seems at the best but dreaming life away."

Sir Walter Scott—a most accomplished rider, and thoroughly versed in all the gaits of the horse—scarcely wrote a poem or a romance without alluding to this motion. In "Red Gauntlet," Letter I., we find: "The black . . . ambled as easily with Sam and the portmanteau as with you and your load of law-learning." And in the same work, Letter XII.: "Better the nag that ambles a' the day, than him that makes a brattle for a mile, and then's dune wi' the road."

The characteristics of the gait are well described in the "Fortunes of Nigel," v.: "He again turned his mule's head westwards, and crossed Temple-Bar at that slow and decent amble which at once became his rank, and civic importance."

Washington Irving, in "Bracebridge Hall," says, "Lady Lillicraft . . rode her sleek ambling pony, whose motion was as easy as a rocking chair."

Cooper, in "The Last of the Mohicans," ii., specifically describes the motion as "a pace between a trot and a walk, and at a rate which kept the sure-footed and peculiar animals they [the ladies] rode, at a fast yet easy amble."

Tennyson, in "The Lady of Shalott," recognizes the suitability of the gait to—

> "An abbot on an ambling pad."

Macaulay, "History of England," iv 25, records the third William as "ambling on a favorite horse named Sorrel" when he met with the accident that cost him his life.

Charles Dickens, in "Barnaby Rudge," xiv., mentions "the grey mare, who breaking from her sober amble into a gentle trot emulated the pace of Edward Chester's horse."

We find in Lord Lytton's "My Novel," iv. 11: "The pad . . giving a petulant whisk of her tail, quickened her amble into a short trot."

Captain Burnaby, in "Horse-back through Asia Minor," xv., testifies that "the pace of a Rahvan, or ambling horse, is an easy one for the rider."

The writer has selected these quotations, from a large number which he has accumulated, for the purpose of clearly illustrating the distinguishing features of a most enjoyable method of riding, which is regrettably so little practised at the present time, except in those parts of the world—Spain, Spanish America, California, and Kentucky, for example—where the gait is better known and its advantages more generally appreciated; and he can emphatically endorse the opinions of the authors quoted, especially those of Cervantes and Irving. Travelling in Central America he has slept for hours at a time while riding—like Prior Aymer—"upon a well-fed ambling mule."

THE AMBLE.

Series 24

A HALF-STRIDE IN SIX PHASES

Horse "Clinton"

Length of complete stride 82 inches (2·05 metres). Time-intervals ·055 second Approximate time of complete stride, ·55 second.

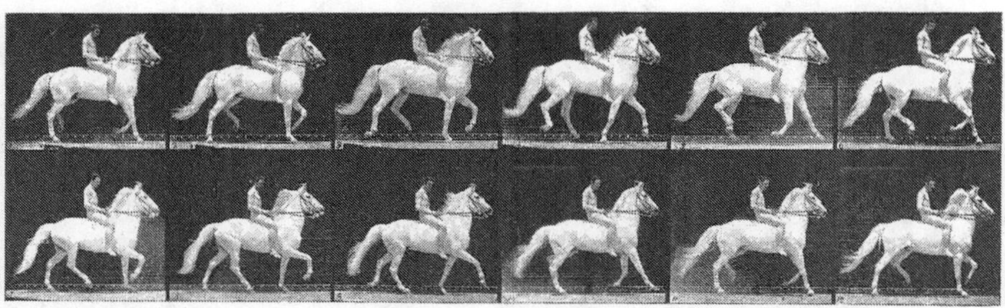

CONTINUATION OF PHASES. SERIES 24.
Horse "Clinton"

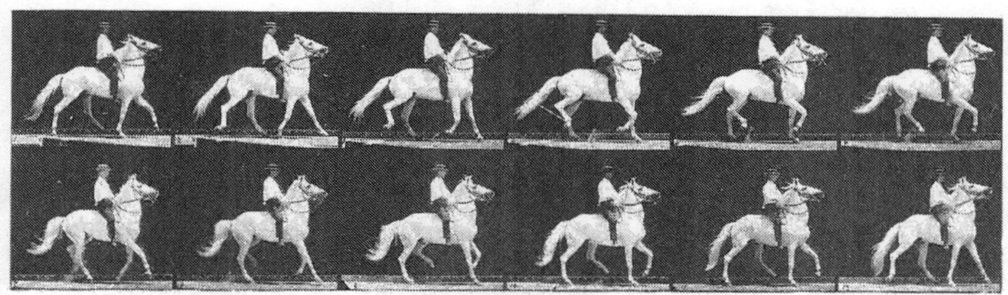

Series 25.

AN IRREGULAR STRIDE IN ELEVEN PHASES.
Horse "Clinton"

Time-intervals ·052 second.

THE AMBLE.

ONE STRIDE IN ELEVEN PHASES

ENLARGED FROM A PHOTOGRAPH ON PAPER. OUTLINES FILLED IN

Horse "Sharon."

Length of stride 123 inches Photographed at Palo Alto, 1879.

THE AMBLE.

Series 21.

ONE STRIDE NEARLY COMPLETED.

The Elephant.

Height of animal. 100 inches (2 50 metres). 452 strides to a mile.

Length of stride: 140 inches (3 50 metres)

Time-intervals: 036 second.

Approximate time of stride ·9 second, or a mile in less than 7 minutes.

THE AMBLE.

SOME PHASES SELECTED FROM SERIES 27.

The Elephant

THE TROT.

SOME PHASES OF THE TROT

The trot is a system of progress in which each pair of diagonal feet are alternately lifted with more or less synchronism, thrust forward, and again placed on the ground; the body of the animal making a transit, without support, twice during each stride.

In this gait there is no inflexible rule as to whether a fore-foot or its diagonal hind-foot, in their respective steps, is first in being lifted and placed on the ground; it is, however, usual for a horse, especially when trotting fast, to give precedence to a fore-foot.

A good example of this gait is given in series 28—a stride by a celebrated trotter, photographed at Palo Alto in 1879.

In phase 1 ● is about to follow the example of the other three feet, and will presently leave the horse without support until 4, when ▲ is found preparing for immediate contact, to be followed without much delay by ○. 5 and 6 show the right fore-leg in a nearly vertical position, with △ elevated nearly to its shoulder. ▲ and ○ render combined support until a period that occurs between 8 and 9, when ▲ is lifted, and leaves ○ exercising its final propulsive force. Two steps, or one-half of the stride, have now been made. The remaining two steps are executed in practically the same manner, the stride is completed in 18, where the limbs occupy almost precisely the same relative positions in which we found them in 1. The two remaining phases commence another stride.

As no chronograph was used in the investigation of 1879, the time made in a mile, or fraction of a mile, was kept by a stop-watch. The stride illustrated was one of three hundred and ten, or thereabouts, made by the horse around a mile track in two minutes and sixteen seconds. A mile has been trotted over in several seconds less time, but the series fairly represents the stride of a first-class trotting horse at the height of his speed.

The length of the stride is readily measured. The lines on the track were twelve inches apart; they show the distance to have been approximately two hundred and four inches, sixty-six of which were made without contact with the ground. Some horses making a stride of not much greater length have been photographed, with the result of showing the transit, without support, to be fully one-half the length of the stride; this, of itself, is, however, no evidence of a more rapid motion than when the feet are on the ground for a longer period.

In the analyzed stride the sequence of phases are—

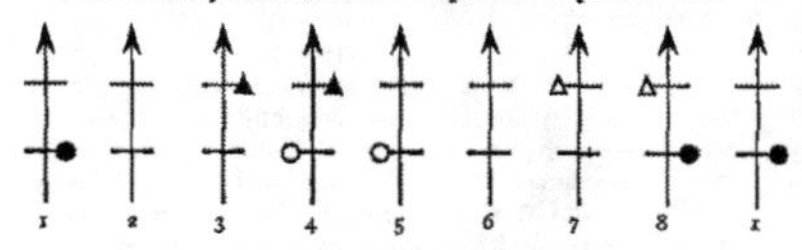

For the purpose of instituting a comparison between the strides of a trot made under different conditions, this same horse was saddled, and went around the track with a jockey on his back. The time was three seconds more, the stride nine inches less, and the distance over which the horse was carried by its momentum, free from contact, was reduced by twenty-four inches.

As the consecutive phases recorded above are not invariably followed, even by the same horse, in consequence, perhaps, of inequalities on the surface of the track, or from some other cause, it will be a safer plan to give a broader significance to a stride of the trot, and to represent it with a more elastic diagram.

For general purposes this definition is perhaps sufficiently exact.

Series 29 is of a good, well-trained horse, going at a moderate speed, with an easy stride.

Series 30 is an example of a stride free from the restrictions of harness or a rider.

Series 33 determines that, *no matter how heavily built* a horse may be, or how slowly he is trotting, the legs relinquish the support of the body twice during each stride; the feet may be merely dragged over the surface, but for a time they are practically inert. This occurs in the trot of all animals; it is demonstrated by the ox, wapiti, eland, fallow-deer, dog, and the cat, in their respective seriates.

On p. 119 are four phases selected from the stride of a high-stepping trotter, which demonstrates in a decidedly pronounced manner the usual sequence of foot-fallings; the high action, however, is not conducive to speed, as much time and labour is wasted in unnecessary exertion.

About a century ago Garrard, an artist of note, painted a picture of the Duke of Hamilton riding a horse, trotting, entirely clear of the ground. The phase seems to have been an innovation that was not acceptable either to other artists or to the public.

We have seen, in the Preface, that so recently as twenty-five years ago, it was the common opinion of those who were supposed to have studied the motion of a horse, that while trotting he always had at least one foot in contact with the ground.

The Romans were familiar with this pace; but as they were accustomed to the amble, they did not appreciate it. They called a trotting horse a "succussator," or shaker; a negative evidence that a racking horse was unknown to them.

References to the trot are frequent in English poetry. Chaucer alludes to it in "The Merchant's Tale;" and Spenser, in "Faerie Queene," iv. 8, says—

> "Whose steadie hand was faine his steede to guyde,
> And all the way from trotting hard to spare
> So was his toyle the more, the more that was his care."

Sir Philip Sidney, "Arcadia," ii.: "I flatly ran away from him toward my horse, who trotting after the company . . ."

The gait was evidently not a favourite one of Shakespeare's; in a metaphor, "As You Like It," iii. 2, Rosalind says, "Time . . . trots hard."

Swift, on the contrary, causes Gulliver, in "A Voyage to the Houyhnhnms," x., to look upon their "gait and gesture . . . with delight," and took it "as a great compliment" when his friends, on his return, told him that he "trotted like a horse."

Scott, in nearly all his romances, speaks of the motion, with high, round, full, hard, reasonable, rapid, stumbling, or other prefix.

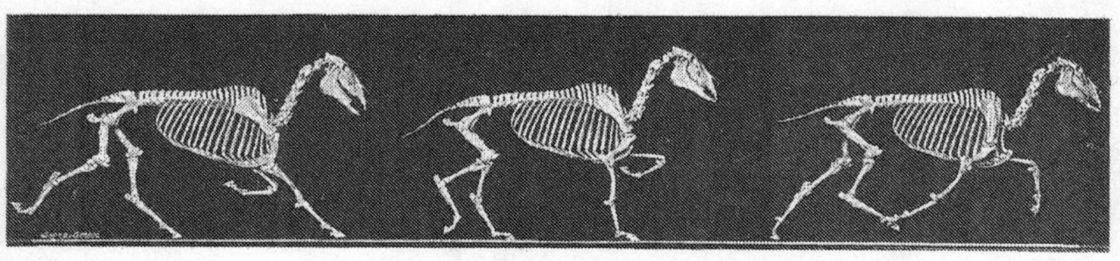

SOME PHASES IN THE FAST TROT OF A HORSE

THE TROT.

Series 28

ONE STRIDE IN EIGHTEEN PHASES.

Enlarged from a Photograph on Paper, Printed in 1879.

Horse "Edgington"

Length of stride: 204 inches (5·25 metres). Free from contact with the ground: 66 inches (1·65 metres).
Approximate time of stride: ·44 second. Strides to a mile: 310

This series was photographed at Palo Alto, 1879, is absolutely free from "retouching," and was synthetically reproduced, and exhibited by projection with the Zoopraxiscope at San Francisco, 1880; at Paris, 1881; and at the Royal Institution and Royal Academy of Arts, London, 1882.

THE TROT.

ONE STRIDE IN EIGHTEEN PHASES

Horse "Daisy"

Length of stride · 118 inches (2.96 metres)

THE TROT.

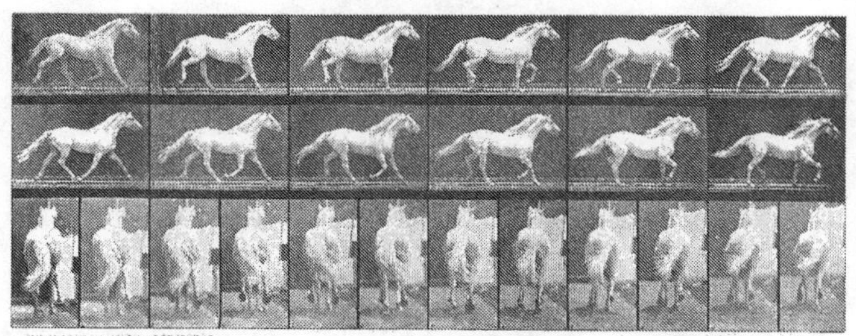

Series 30.

A HALF-STRIDE IN NINE PHASES, PHOTOGRAPHED SYNCHRONOUSLY FROM TWO POINTS OF VIEW.

Horse "Eagle"

Length of complete stride: 163 inches (4·12 metres) Time-intervals ·045 second

Approximate time of complete stride, ·65 second.

THE TROT.

SOME PHASES OF THE TROT FROM SERIES 30.

THE TROT.

Series 31.

ONE STRIDE, PHOTOGRAPHED SYNCHRONOUSLY FROM TWO POINTS OF VIEW.

Horse "Beauty"

Length of stride 112 inches (2·80 metres). Time-intervals ·052 second Approximate time of stride ·55 second

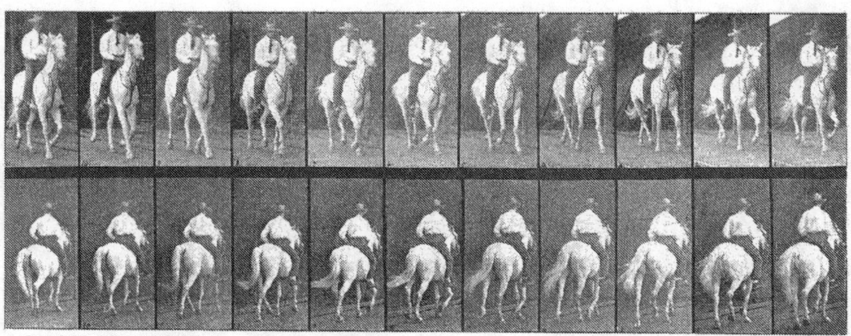

SERIES 32

ONE STRIDE, PHOTOGRAPHED SYNCHRONOUSLY FROM TWO POINTS OF VIEW
Horse "Elberon."

Time-intervals 046 second. Approximate time of stride ·46 second.

SERIES 33.

A HALF-STRIDE IN NINE PHASES, PHOTOGRAPHED SYNCHRONOUSLY FROM TWO POINTS OF VIEW
Horse "Dusel."

Time-intervals · 056 second. Approximate time of stride 95 second.

THE TROT.

SOME PHASES IN THE MOTION OF A HORSE TROTTING AT A HIGH RATE OF SPEED.

SOME PHASES IN THE MOTION OF A HORSE TROTTING SLOWLY

THE TROT.

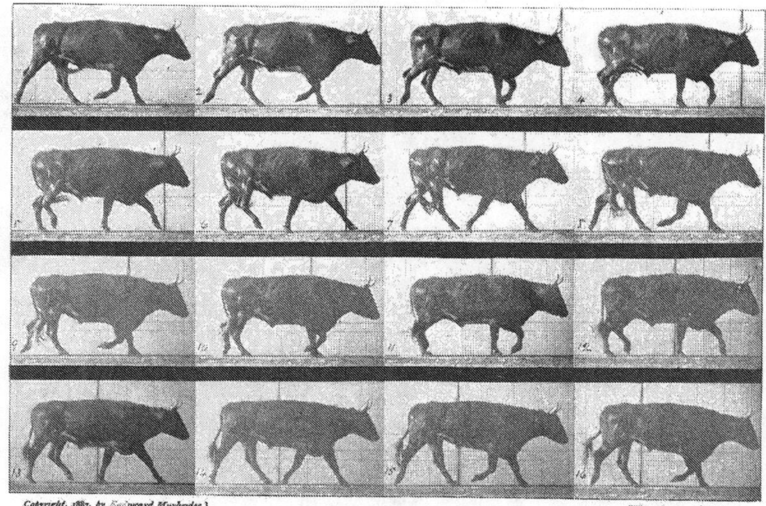

Copyright, 1887, by Eadweard Muybridge.

SERIES 34

ONE STRIDE IN FIFTEEN PHASES.

The Ox

Length of stride 104 inches (2 60 metres) — Time-intervals 056 second — Approximate time of stride. 78 second

THE TROT.

A HALF-STRIDE IN THIRTEEN PHASES.

The Wapiti, or Elk.

Length of complete stride 148 inches (3.50 metres) Time-intervals ·032 second. Approximate time of complete stride 73 second.

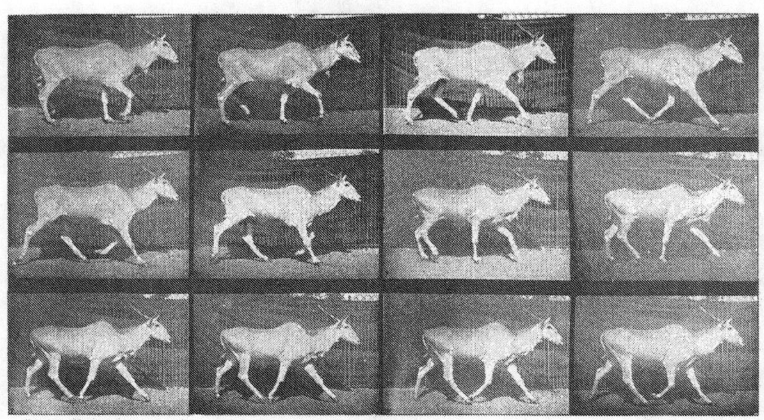

 Series 36

A HALF-STRIDE IN UNEQUAL PHASES.

The Eland.

THE TROT.

 ← SERIES 37

ONE STRIDE IN TEN PHASES.

The Fallow Deer

Time-intervals: 069 second. Approximate time of stride 62 second.

THE TROT.

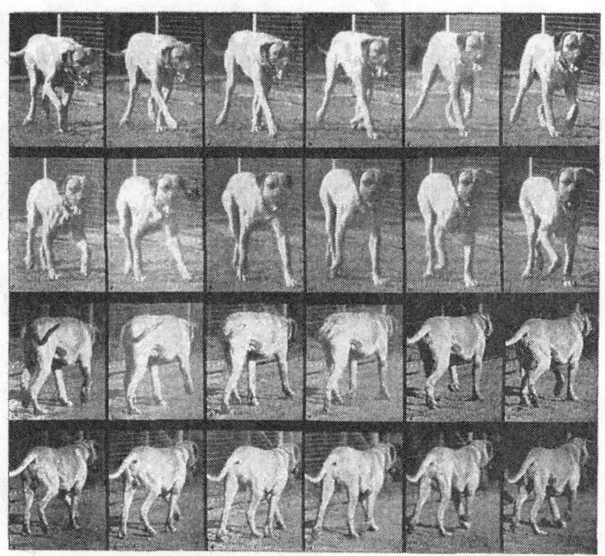

 SERIES 38

ONE STRIDE, PHOTOGRAPHED SYNCHRONOUSLY FROM TWO POINTS OF VIEW

The Dog (Mastiff)

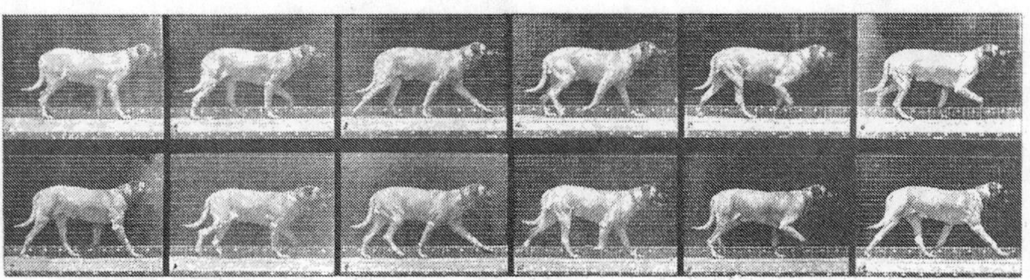

 Series 39

AN IRREGULAR STRIDE

The Dog (Mastiff)

Length of stride 52 inches (1·30 metres) Time-intervals: ·110 second

Approximate time of completed movement: 1·21 seconds

 Series 40.

ONE STRIDE IN TEN PHASES.

The Cat

Time-intervals: ·035 second.

THE RACK.

In the rack, the legs of the animal are used in lateral pairs, instead of, as in the trot, diagonal pairs. The same uncertainty with regard to precedence of the fore or hind foot-impacts prevails in this gait, as in the trot; in contradistinction to the latter, priority is usually given in the rack to a hind-foot, this being so immediately followed by its lateral fore that, practically, they may be said to swing simultaneously.

This being an awkward, and, to the rider, an exceedingly disagreeable method of locomotion, horses are, happily, rarely trained to its use; when they are, it is for traction, in the expectation of gaining some slight advantage in point of time, over the trot.

A profile silhouette picture of any phase of the rack would be indistinguishable from a phase of the trot.

Series 41 is representative of an average stride during a moderately fast rate of speed. In 1 the horse has just alighted on ○; △ is preparing to follow. 2 shows both legs nearly vertical; and 4, a transit without support, ▲ somewhat elevated, and ● skimming over the surface. In 5 the right laterals have assumed the functions of support, which on the evidence of the pasterns was commenced by ●. Two steps, or one-half the stride, are now finished; the remaining phases lead to the discovery of the following two steps having been completed in practically the same manner. A diagram of this stride may be therefore shown as—

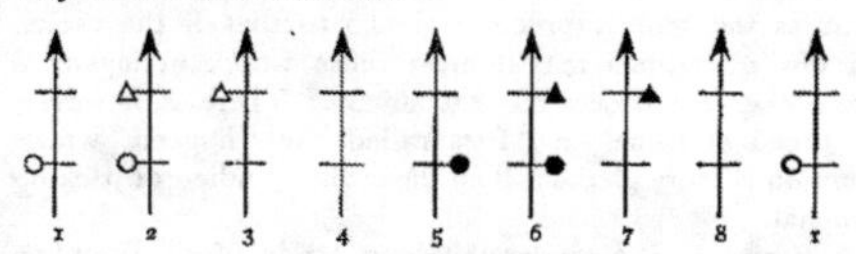

Taking into consideration that precedence is not invariably given to a hind-foot; a stride of the rack, for general purposes, may be represented as—

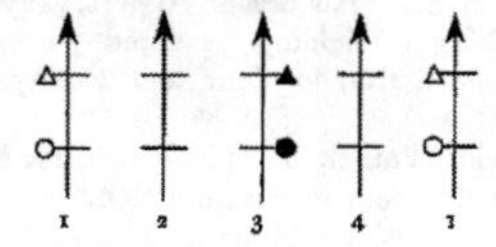

The rack is a gait natural to the camel, the giraffe, and some few other animals; it is said to have been occasionally observed in the dog.

If the horse was ever trained to rack for the use of a rider, it was probably for some one who wished to subject himself to a penance; the personal experience of those who have tried it, induces an imaginary comparison with the torture which, a few centuries ago, it was the custom to inflict on recalcitrants with an instrument from which the gait probably takes its name.

No references have been found descriptive of the sensations experienced by the rider of a racking horse; but as the gait is precisely similar to that of the camel, a few quotations may interest those who contemplate a ride over the desert on that animal. There is, however, a breed of camel in Africa called the "hygeen," whose motion is more pleasant than the ordinary riding or packing animal.

Series 42 is a moderately long stride of an Egyptian camel. As with other animals, long confinement had impaired its capability of speed. For artistic purposes the motion is well represented.

Morgan, in his "History of Algiers," says the camel "makes nothing of holding its rapid pace, which is a most violent hard trot, for four and twenty hours at a stretch."

Beckford, in "Vathek": "The rough trot of Alboufaki [a camel] awoke them in consternation."

G. W. Curtis, "The Howadji in Syria": "The trot of the usual travelling camel is very hard . . . but MacWhirter's [his own camel's] exertions in that kind shook my soul within me."

It will be observed that each of these travellers speaks of the camel's gait as a "trot." The author can find no evidence of a camel ever having been trained to trot, it certainly is not its natural gait.

In the sixteenth century, George Peele, in an Eclogue, says—

> "His Rain-deer racking with proud and stately pace
> Giveth to his flock a right beautiful grace"

The application of "racking" to the pace of the rein-deer seems to require some explanation. That animal, like other deer, trots; and no trotting animal racks naturally.

This system of motion, under the illogical name of "pace," has, mysteriously, been confounded with its very antithesis of gaits—the amble. Why a name applicable, in its broad sense, to motion of any kind, should ever have been allotted to a special method of animal progress, is a question that defies elucidation. The absurdity of its use as a distinctive gait is self-evident.

Dante (Cary), in "Purgatory," xxiv., has—

> "And as a man
> Tired with the motion of a trotting steed
> Slacks pace, and stays behind . . ."

Scott, in "Rob Roy," iii., remarks, "The trot is the true pace for the hackney."

"Guy Mannering," xiii.: "Dumple, quickening his pace, trotted about a mile."

"Red Gauntlet," Letter VI. "The rider . . . slackened his horse's pace from a slow trot to a walk."

And in the "Ingoldsby Legends" (the Execution) we find—

> "Adown Piccadilly and Waterloo-place,
> Went the high-trotting mare at a very quick pace"

The designs which seem to indicate the rack on Etruscan, Greek, and Roman vases are probably due to artistic indifference. It is an unnecessary and unnatural gait of the horse, and it is scarcely probable that the ancients trained the animal to its use.

THE RACK.

SERIES 41.

ONE STRIDE IN NINE PHASES, PHOTOGRAPHED SYNCHRONOUSLY FROM TWO POINTS OF VIEW.

Horse "Pronto"

Length of stride: 146 inches (3·70 metres). Time-intervals ·076 second. Approximate time of stride ·64 second.

SERIES 40.

A HALF-STRIDE IN FIFTEEN PHASES.

The Egyptian Camel

Length of complete stride · 146 inches (3·70 metres) Time-intervals ·024 second

Approximate time of complete stride ·65 second.

THE RACK.

The Egyptian Camel.

SOME PHASES OF THE RACK.

Horse "Pronto"

THE CANTER.

We have hitherto devoted our attention to systems of locomotion which permit the division of a stride into two co-ordinate parts, each of which, with a reciprocation of limb action, is essentially a repetition of the other.

We now come to a different class of motion, the strides of which cannot be so divided, and each one must be considered as a unit, unsuited for equitable partition.

The canter has the same sequence of foot-fallings as the walk, but without the same regularity of intervals, and during a portion of the stride the body has a longer or shorter unsupported transit. In this gait the spring is invariably taken from a fore-foot, while the landing is effected on the diagonal hind-foot.

Series 43 demonstrates the spring on the point of being taken by △ in 4; ● is not squarely on the ground until phase 9 is attained; the other three feet in the meanwhile are being gradually thrust forward. In 12 ▲ comes to the assistance of ●, when the support is administered by the right laterals, but for a very brief period; ○ quickly follows in the wake of its diagonal, and in the next phase the rear part of its shoe is in close proximity to the ground. So rapidly does the following usually take place, that the ear is frequently incapable of recognizing an interval between the successive sounds of the foot-impacts. We now find ● ▲ ○ engaged in supporting the body, ▲ having the greatest strain.

At an ordinary speed the first hind-foot to fall is lifted in advance of the second fore-foot's descent, and, as in 16 and 17, the diagonals assume the responsibility of support. △ is now brought to the relief of ▲ ○; the former, however, soon dissolves the tripartite, and relinquishes its offices in favour of the left laterals; this partnership is of brief duration, for in 22 we find ○ deserting its post, and leaving △ to its solitary labours, which it satisfactorily performs through several phases, when it joins its companions in the enjoyment of a period of rest, from which ● will again be the first to go to work.

From this analysis we ascertain the sequence of phases in a representative stride to be—

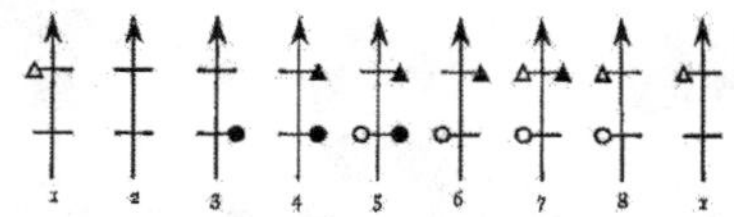

Had the spring been made from ▲ the landing would, of course, have been made on ○, the others falling in regular order

During a slow canter △ will sometimes be discovered acting in association with the other three, and the curious phase presented of a five-mile gait being realized, with all four of the feet in contact with the ground at the same instant.

A PHASE OF THE SLOW CANTER.

The earliest reference to the canter in English literature is probably in a seventeenth-century book by Brathwait, "Clitus' Whimsies," who alludes to the gait as "a Canterbury."

Dr. Thomas Sheridan, in a poetical letter to Swift, says—

"When your Pegasus cantered in triple and rid fast."

Dennis, "On the Preliminary to the Dunciad," has: "The Pegasus of Pope, like a Kentish post-horse, is always on the Canterbury."

Burns, in "Tam Samson's Elegy," and in "Tam o'Shanter," refers to the "canter" simply; as does also Combe, in "Dr. Syntax," xxvii. and xxxviii.

Scott, in "Waverley," xv., "St. Ronan's Well," i., and "Guy Mannering," xxiii., writes of "cantering," "easy canter," and "cantered." In "Talisman," xxii., "Old Mortality," xliv., and "Red Gauntlet," Letter IV., he alludes to the same gait as "a hand-gallop."

Byron, Fennimore Cooper, Washington Irving, Tennyson, Lytton, Dickens, Smedley, Dobson, Darley, Sir F. H. Doyle, Charles Reade, Dr. Livingstone, Saxe, and many other authors, in prose and poetry, discard the "bury," and advert to the motion simply as a "canter."

The notion of this gait of the horse deriving its name from its association with the "Canterbury Pilgrims" is untenable. Many passages in the "Tales," and the illuminations in the Ellesmere manuscript, disprove the supposition, as we have already seen in the "amble."

It is far more likely that the alternate rising and falling of the fore- and hind-quarters of a horse in the execution of the movement, suggested a resemblance to the alternate tilting or "canting" of a plank on which children sit in the game called "see-saw." If a saddle were to be arranged over the fulcrum, and the plank rapidly but gently "canted" up and down, a rider on the saddle would not fail to experience a sensation similar to that produced by the canter of a horse.

THE CANTER.

Series 43.

ONE STRIDE IN TWENTY-THREE PHASES

Horse "Daisy."

Length of stride 96 inches (2·40 metres) Approximate time of stride: 60 second.

THE CANTER.

Series 44

ONE STRIDE IN TEN PHASES.

Horse "Clinton."

Length of stride, 108 inches (2.70 metres). Approximate time of stride, .52 second.

THE CANTER.

SOME PHASES OF THE CANTER FROM SERIES 44

Horse "Clinton"

THE CANTER.

Series 45.

ONE STRIDE PHOTOGRAPHED SYNCHRONOUSLY FROM TWO POINTS OF VIEW

Thorough-bred Mare "Annie."

Length of stride: 114 inches

THE GALLOP.

The word "gallop," in its various forms of spelling, is now almost universally employed to designate the most rapid of all quadrupedal movements. The action is adopted by nearly all animals, in one or the other of its methods, when, from caprice, persuasion, or necessity, they exercise their utmost power for the attainment of their greatest speed.

Photographic analysis demonstrates two systems of galloping; one in which the foot-impacts individually succeed each other in a way that may be conveniently represented by the points of a cross—

TRANSVERSE-GALLOP.

the other, in which the limb movements and consequent foot-fallings succeed each other in a rotative manner, which may be roughly represented by a circle—

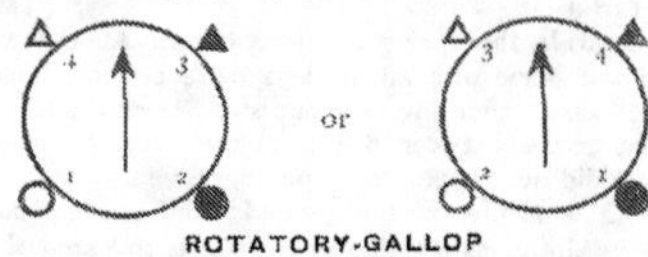

ROTATORY-GALLOP

In these diagrams, the notation commences with the fall of a hind-foot on the ground after an unsupported transit of the body.

To these two systems of galloping, the names of "transverse-gallop" and "rotatory-gallop" may appropriately be given; if they are too cumbersome for popular usage, the prefixes "cross" and "rota" respectively can, perhaps, be correctly applied. "Round" might have been used in association with the latter, but the word is already in use to imply a rapid progress by any gait

The transverse-gallop is employed by the horse, and by the greater number of other animals, both horny and soft-footed, the rotatory-gallop is adopted by the dog, the deer, and some other animals.

We will devote our attention, firstly, to the transverse or cross-gallop.

Series 46 illustrates twenty-one consecutive phases, which occurred in one stride of a thorough-bred Kentucky horse, exerting all his power to gallop at his highest speed.

For convenience of reference, the analysis commences with phase 2, and is completed with 22, although it will be noticed an inch or so more of progress is necessary to obtain an exactly corresponding phase to that with which we commence; and for the greater convenience of the student it will be sufficient to assume that the distance-intervals of the phases are thirteen and three-quarter inches. The time-intervals, as recorded by the chronograph, are exactly twenty-two one-thousandths of a second each.

In this stride, the spring is effected from ▲, and we soon find the horse with all his legs more or less flexed under the body, affording no support thereto until a period that occurs between 6 and 7; the exact phase of first contact did not happen to be photographed.

In 7 ○ is firmly on the ground; the pastern has exercised its duties as a spring or a cushion to lessen the concussion; the heel being already in close proximity to the ground, into which it is impressed in 8 and 9. The distance the body was hurled through the air, with the final assistance of ▲, was about seventy-eight inches in a little more than the tenth of a second.

In 10 ● has just commenced assisting ○; they do not, however, long remain in company, for in 12 ○ is already lifted, and upon ● devolves the unaided duty of support. ○ and ● present the shortest distance-interval of combined support, forty-six inches only; 13 discloses △ in actual contact with the ground, but the pastern has not yet commenced to bend; support is now furnished by the diagonals at a distance of ninety inches from each other. As the leg of △ becomes vertical, the pastern gradually becomes horizontal, until in 14 and 15 its joint is impressed into the ground. The great weight of the horse commences in 14 to be thrown on △, which receives no assistance until a little beyond 17, in which phase the shoe of ▲ is yet about two inches above the track. The combined support of △ and ▲ is of very brief duration, for the great distance they are apart (sixty inches) renders much progress, without separation, impossible.

In 18 ▲ has been on the ground for a considerable time, as is demonstrated by the nearly vertical position of the leg, and the consequent bending of the pastern.

It is interesting to note the enormous amount of work this leg has to do, for in its duplex offices of support and propulsion it receives no assistance through eight intervals, a distance of much more than one hundred inches. If each of the legs of the horse had carried him an equal distance during this stride, it would have measured more than twelve yards.

For the purpose of giving a concise demonstration of the stride of a first-class thorough-bred horse, in fine condition and in good training, over a well-kept racing-track, some of the phases of series 46 are reproduced, and enlarged, on p. 173 The nine phases show seven different methods of support, and a period of unsupported transit. They are, as in all orthodox strides of the transverse-gallop, when the spring is made from ▲—

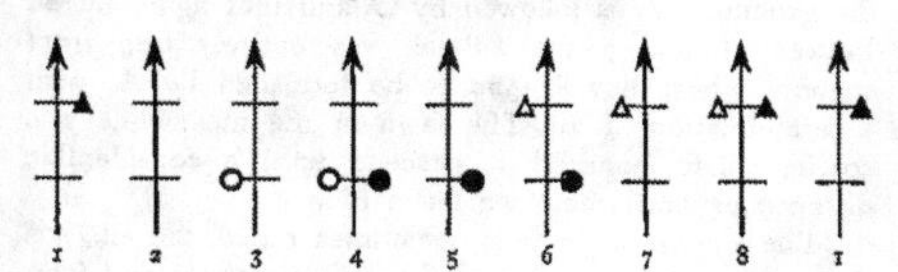

Had the spring been made from △, the landing would have taken place on ●, and corresponding alterations would have been made in the sequence of the other foot-impacts.

This stride is not, of course, presented as a record-breaker—longer strides are frequently made, and a mile galloped over in less proportionate time—but it may be accepted as a fair average stride of a first-class thorough-bred horse, made during a race with equally good competitors, a second or so before reaching the winning-post.

Series 47 is the stride of a thorough-bred mare; it is interesting for comparison with others.

In seriates 48 and 49, the horses are not squarely on a regular gait, having had to swerve from a straight course to permit synchronous fore-shortened phases to be made.

The ineffectual attempt of a heavily built draught-horse to emulate the speed of a thorough-bred when, with the same succession of impacts, a hind-foot or a fore-foot is sometimes flat on the ground in association with two other feet, as in series 51, results in a gait which may be called an irregular, or abnormal gallop.

The transverse succession of foot-fallings is found in the gallop of the buffalo, the goat, the camel, and the cat, as illustrated by their respective seriates. The latter animal, having a greater flexibility of movement, combines with the orthodox stride a spring into the air from its hind feet; the foot-impacts, however, have the same sequence as those of the bear, raccoon, and hog, as demonstrated on p. 201.

It is probable that future research will discover—with the horse and some other animals—during extreme speed, an unsupported transit from one anterior foot to the other.

Some writers claim for the horse a more rapid gait than that of galloping, to which they have given the name of "running." It is definitely proved that the rapid gallop of the horse is executed in one way only; at present he has no faster gait. In its reference to quadrupedal movements, "running" can be applied only as it is to a stream of water running down a hill, a locomotive running along a railroad, or an ivy plant running up a wall.

Whether it would be possible to obtain a higher rate of hereditary speed, after horses had been taught to practise the rota instead of the cross-gallop, is problematical.

In the rotatory or rota-gallop, a different system of foot-fallings prevails; the consecutive supports revolve, as it were, in one or the other direction around the body of the animal. This method of galloping is satisfactorily demonstrated by series 56, two strides of a small coursing hound, with a national reputation for speed, and, although only about 16 inches high, the winner of many trophies from larger animals.

We will commence the analysis with phase 6, which exhibits the hound on ▲, and about to spring therefrom into the air, where we find him in 7, with all the legs flexed under the body; the two fore feet far to the rear of ●, on which he presently alights, and quickly follows with O, from which he takes another spring, and in 9 we find a phase somewhat resembling the unique modern conventionality of the galloping horse. After a flight with outstretched legs, the landing takes place on △. In 11 the support is transferred to ▲, and we arrive at a virtual repetition of the phase with which we commenced in 6.

In series 57 a powerful, heavily built mastiff is doing his best to emulate the speed of the racing hound; his weight, however, is against him, and although he effects a spring from a fore-foot, it is beyond his capability to spring from a hind-foot.

Owing to the extreme heat of the day, the manipulation of series 58 leaves much to be desired; but it is a good illustration of the stride of a fallow-deer in captivity, followed by its frightened fawn.

In phase 2 ▲ is on the ground, and is followed by △; had not long confinement in a small park impaired the elasticity which the deer would have exhibited in its natural state, a phase would have occurred between those of 3 and 4, in which all the feet would have been off the ground. △ is followed by O, and that again by ●. Between 6 and 7 the animal was entirely free from support, which now begins to be furnished by ▲, with a recapitulation of 2. The fawn in the meanwhile was soaring aloft, nor did it descend until a considerable distance beyond where we leave it in 9.

The wapiti, or, as it is sometimes called, the elk, has the same rotative sequence of limb movements and foot-impacts as the dog and the deer; so also has the antelope, and it is probable it will be found with the moose.

The diagram reproduces at a glance the sequential phases of the rota-gallop; or, the rotation may be accomplished in the reverse direction.

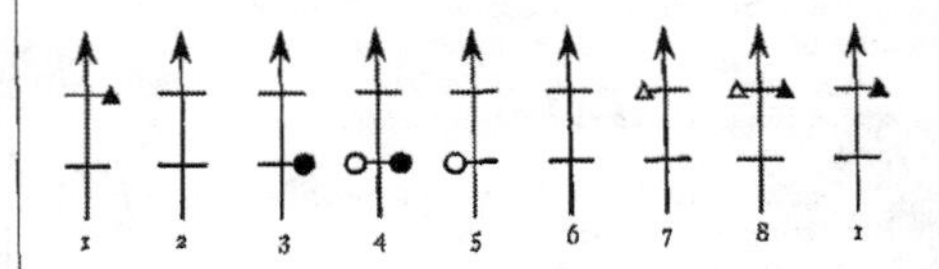

SILHOUETTES SELECTED FROM THE RESULTS OF THE PALO ALTO INVESTIGATION, 1872–79, ALL OF WHICH RESEMBLE PHASES THAT HAVE BEEN, AT VARIOUS TIMES, ADOPTED BY ARTISTS AS THEIR INTERPRETATION OF THE GALLOP OF THE HORSE

A history of the artistic delineation of the gallop is worthy of attention; it is hoped, one will some day be written, and comprehensively illustrated.

Pending its appearance, it is impossible, in this volume, to do more than casually refer to the expressions given to this method of locomotion by the artists of a few nations, at different epochs, as represented by the horse, and in general terms to consider their predominant characteristics.

With this object only in view it will be sufficient to arrange the prevailing traits of its treatment in three broad classifications.

First, the Primitive; suggested to the artist by keen observation, and expressed by him with entire freedom from conventionality.

The distinguishing features of this type are the flexure of all the legs more or less under the body, with one or both the hind feet free from contact with the ground.

Examples from nature, the impression of which influenced, as it continues to influence, the untaught and the unconventional artist, may be found in figures 1 and 2 of the line of silhouettes, phases 5, 6, and 7, series 46; and 1 to 4, series 52.

Second, the Ancient. In which the support is rendered by the two hind feet, the anterior legs are more or less flexed, with their feet in close proximity, and raised at various elevations above the ground. Figures 3 and 5 of the silhouettes; phase 6, series 64; and 2 of series 66, resemble this interpretation.

Third, the Modern. Which, so far as it is used in depicting a regular progressive motion of the horse, exhibits an entire absence of careful observation, unprejudiced impression, or serious reasoning.

In its most pronounced realization it is characterized by a body, neck, and head, all of abnormal length, and arranged in a nearly horizontal line. The anterior legs, in almost parallel lines, have their feet a few inches below, and in advance of, the nose. Both hind feet are thrust far to the rear, with their shoes turned upwards

There is no phase in the motion of a horse photographed from life that can be referred to as an example of this curious treatment of the gallop, nor will any combination of phases in the motion of the animal convey an impression resembling it. A suggestion of it may be found in figure 9 of the silhouettes, but it may perhaps be more correctly represented by phase 16, p. 225, which occurs in the leap of a cat.

The segregation of the innumerable different representations of the gallop into three principal groups is, of course, purely arbitrary. Each group is susceptible of subdivision, especially that classified as Ancient, of which certain differences may be noted in the Egyptian, the Assyrian, the Grecian, the Roman, and the Byzantine method of treatment. The grouping is merely intended to indicate the general idea which seems to have influenced the artist when his evident intention was to represent the animal at full speed, in accordance with the prevailing fashion of his nation or his time.

It is as well, perhaps, to remark, that in the line of silhouettes, figures 1, 2, and 9 only, are phases which occur in regular progressive motion; 3 and 4 are selected from seriates of preparations for a leap over a hurdle; 7 and 8 represent the recovery from the first contact with the ground after a leap; 5 and 6 illustrate phases of the capering which a horse sometimes indulges in before starting on a regular method of progress.

To the student who wishes to inquire more minutely into the history of the artistic gallop, the following references to some examples of its treatment may be found useful.

In the Museum of Art at New York is a well-preserved porphyry cylinder, unearthed by the archæologist Ward on the plains of Chaldæa, and pronounced to be of Hittite manufacture, of date probably 3000 to 2000 B.C.

Among other designs thereon are the figures of two men with their arms upraised, stampeding a herd of cattle, which are evidently fleeing from them at their utmost speed. The animals are represented with all their legs flexed under their bodies, somewhat like those of the buffalo (2, 3, and 4, series 52).

In the "Skandinaviens Hallristningar Arkeologisk Afhandling," 1848, Dr. Holmberg reproduces some prehistoric sculptures on a rock at Tegneby, Sweden. Among other figures are several horses, and two groups of horsemen, charging apparently in battle. These designs are probably of earlier date than any others yet found in Northern Europe, and they represent the animals with their legs flexed under their bodies.

On an archaic Mycenæan vase, reproduced in the "Journal of Hellenic Studies," vol. vii., are some figures of a horned animal resembling an ibex, the legs of which are arranged in the same manner.

More recently, the Alaskans, in their etchings on ivory, and, further south, other North American Indians, on their painted buffalo-skins, were accustomed to the use of similar phases as an indication of speed.

An intelligent child, known to the author, who, having a talent for drawing, and, happily, not familiar with the conventional representation of the gallop, was asked to sketch her idea of a runaway horse, which she had seen, produced a similar phase as her impression of the action.

A reference to phases 2 to 6, series 46, demonstrates

that during this portion of the stride of a horse at full gallop, all four of the legs are flexed, and their feet in close proximity, especially in 3, 4, and 5, where they remain without much independent action for a comparatively long period of time, with the result that this class of phase has a stronger and more lasting effect on the retina of the eye than any other class, and conveys to the unprejudiced and to the unsophisticated observer an impression of extreme rapidity.

The horse does not make its appearance in Egyptian art until about 1500 B.C., or shortly before the Israelite exodus. For many centuries it seems to have been used for no other purpose than to drag a chariot in warfare or in a triumphal procession. No evidence of its use for riding purposes appears for nearly a thousand years later.

The battles of Seti and of his son Rameses, carved on the walls of Karnak and other temples, include numbers of chariots, each drawn by two horses in the conventional phase prescribed by law. The probability of this phase—which represents an incident of the leap—having been chosen as an emblem of the triumphant monarch surmounting every obstacle that interposed itself between him and victory, may be worthy of consideration. Whatever its origin, the remarkable fact is disclosed, that this phase of an entirely different and accidental motion of the horse was accepted, with sometimes the modifications introduced by the Greeks, the Romans, or the Byzantines, almost universally, as the symbol of the gallop, by generation after generation of artists for more than thirty centuries.

In the twelfth century B.C. it appears on a slab of marble, discovered at Mycenæ, representing a warrior in a chariot drawn by two horses.

The explorations of Layard prove that in the eighth century B.C. the horse was used by the Assyrians, both for dragging chariots and for riding; several bas-reliefs in the British Museum illustrate the animal being used for these purposes, and with the same indications of rapid motion as those prevailing on the banks of the Nile.

In the gallery of the Louvre is a slab of this period from the temple at Assos, representing a centaur galloping in the same manner. Babylonian coins indicate a similar treatment of the movement.

Even in the caricatures of these times, no other phase seems to have been thought of. In a papyrus, depicting a battle between cats and rats, the animals in the chariots of the attacking parties are, virtually, copies of the horses at Karnak; and a Phœnician vase in the British Museum exhibits in a significant manner the defeat of an Egyptian warrior, who is launching a farewell arrow to the rear, while a solitary horse in his chariot, with anterior feet high in the air, is being driven, with presumed rapidity, homewards.

A Greek gem of the sixth century B.C., has a beautifully executed intaglio engraving of a winged goddess in a chariot, driving two horses in a slightly modified style to that of the Egyptians.

The Nereid monument, now in the British Museum, originally erected at Xanthos, Lycia, four centuries B.C., has a number of horses and dogs engaged in the chase, all of which, with some slight alterations in the fore legs of the animals, are treated in the same manner.

Although the Greek artists, even at the zenith of their fame, frequently represented the gallop in accordance with its borrowed interpretation, their conceptions of the fast motion of a horse were not always restrained by the traditions of the past Many works of art may be found in which rapidity of movement is expressed in phases which exhibit close attention to natural law.

Their teachings soon began to exercise a salutary influence among the artists of other nations with whom they had communication. A curious instance of this may be seen on a silver dish of Phœnician manufacture, discovered by di Cesnola, on the island of Cyprus.

On the border are two horses attached to a chariot, and represented in the orthodox Egyptian fashion; immediately in front are two men, apparently Assyrians, riding horses, in which Greek treatment of a phase occurring in the gallop is very evident.

The Phœnicians were not a creative or art-originating people, and their designs of moving animals seem to have been copied at random, according to the tastes of their patrons.

In the Panathenaic procession there are not any horses to which the action of the gallop was intended to be given. Some appear anxious to start off at a rapid rate, others suggest a sudden check from a fast motion, but none are making a sustained rapid progress.

The proportions of many of these horses are susceptible of criticism.

During the third century B.C., horses with riders began to appear in Egyptian designs, and a phase of motion is used which exhibits an innovation probably due to Greek influence. The earliest gold coin used by the British is of the second century B.C. It has for its obverse a warrior seated on a horse, the motion of which is more suggestive of a later treatment of the gallop by the Byzantines, than it is of that generally adopted either by the earlier Greeks or by the Romans. It is of similar design to a coin of the Macedonian Philip, and was probably stamped in Greece. Some of the British coins of the time of Boadicea bear the effigy of a singularly disjointed horse, undoubtedly a home-manufactured copy of the design on the phillipus.

The Roman modification of the gallop can be advantageously studied from the designs on the column of Trajan, and on the arch of Titus. The extraordinary projection of the fore legs of many ancient sculptures of horses, of which the Biga at Rome is a well-known example, is worthy of attention.

The gallop of the Byzantine Greeks had, probably, its best illustrations on the column of Theodosius, at Constantinople. Some of the horses have a resemblance to figures 5, 6, and 7 of the line of silhouettes.

In the British Museum is a pilaster from the Tope at Amravati, India, carved in the sixth century, which has the front portions of several horses and other animals, galloping in compliance with Egyptian rules; and a number of elaborate mosaics, of the same period, from Carthage, in which horses, hounds, stags, lions, hares, gazelles, and wild boars are represented in hunting scenes, in accordance with the same standard.

If the illustrations in Porter's work faithfully represent the carvings on a Persian temple of the same century, at Tackt-i-Bostan, the designer of some reliefs representing a horse, a deer, and a wild boar seems to have anticipated the interpretations of the nineteenth-century artists.

The ruins of a temple at Angkor Wat, on the borders of Cambodia and Siam, built probably in the ninth century, by a race of people called the Khmers, exhibit carvings on stone of several chariots occupied by Mongolians, and drawn by horses which, in their expression of motion, have a striking resemblance to some of those on the Parthenon frieze

The outlines of a gigantic horse, cut into the side of a Berkshire hill, in supposed commemoration of a victory by King Alfred over the Danes, presents the appearance of having been copied from a Byzantine design.

European artists of this epoch, in their interpretation of the gallop, seem to have lapsed into the original conventionality. Evidence of this is seen in the miniatures of a ninth-century Bible in the Vatican library; in the Anglo-Saxon manuscripts of Prudentius, and of other writers in the Greek and Latin tongues; and it appears with many other known and unknown motions on the Bayeux tapestry of the eleventh century.

In the fourteenth century a remarkable exception to the rule was painted—probably by Pisano—on the walls of the Campo-Santo at Pisa. It is of a mounted knight, whose horse is represented in a phase almost exactly corresponding with that of 9, series 46—a truthful, if not a judicious indication of the gallop

In the miniatures of Froissart may be found the horses at Karnak; which also served as models to some illuminations of the "Canterbury Tales," in the Ellesmere Collection.

Raffaelle, Titian, and many other Italian artists inclined toward the Byzantine modification; not so, however, did their countryman Guido, nor Albert Durer, nor the greater number of German, Dutch, and other artists of the fifteenth, sixteenth, and seventeenth centuries, who for their interpretations resorted, without compunction, to the monotonous designs of their predecessors, whose mummies had been deposited on the banks of the Nile three thousand years before their copyists were born.

Their rendition of the motion was endorsed by the mathematician Borelli, and by the veterinarian Newcastle; so this ancient symbol of a conquering hero having been adopted as an emblem of the gallop, continued to be its one unvarying sign until it disappeared with the eighteenth century.

About a hundred years ago, the artists of Europe, apparently with one accord, came to the conclusion that the rising body, with the bent, uplifted anteriors, and the contact of the hind feet with the ground, as indulged in by the ancient sculptors, was inconsistent with the correct interpretation of speed, and, as if by preconcerted agreement, there suddenly appeared from their various schools the conventional phase which attained the zenith of its absurdity in a well-known picture, by a celebrated animal painter, representing ten horses, each a replica of the other, with limbs extended fore-and-aft, and gliding through the air, distinguishable from each other only by the colours of their riders.

Bearing in mind the axiom of Leonardo da Vinci, one may well suppose the picture was painted with the same object in view as that with which Don Quixote was written.

"And yet," it is sometimes remarked, "the phase gives one an impression of rapid motion." Possibly, but in precisely the same way as a printed word unconsciously suggests, through long usage, the sound or the substance of that which it represents.

If it is impressed on our minds in infancy that a certain arbitrary symbol indicates an existing fact; if this same association of emblem and reality is reiterated at the preparatory school, insisted upon at college, and pronounced correct at the university; symbol and fact—or supposed fact—become so intimately blended that it is extremely difficult to disassociate them, even when reason and personal observation teaches us they have no true relationship.

So it is with the conventional galloping horse; we have become so accustomed to see it in art that it has imperceptibly dominated our understanding, and we think the representation to be unimpeachable, until we throw all our preconceived impressions on one side, and seek the truth by independent observation from Nature herself.

During the past few years the artist has become convinced that this definition of the horse's gallop does not harmonize with his own unbiased impression, and he is making rapid progress in his efforts to sweep away prejudice, and effect the complete reform that is gradually but surely coming.

A PHASE IN THE GALLOP OF THE HORSE.

Many art designs, both ancient and modern, represent a horse performing some feat of locomotion, with not merely that portion of the anterior limb technically called the "fore-arm," but even the "elbow," thrust forward beyond the nose.

For the purpose of ascertaining how far it was possible for a horse, during regular progress, to extend his fore-foot, a thorough-bred Kentuckian, noted for his long stride, was selected for an experiment at Palo Alto in 1879.

The line of silhouettes represents a single phase of motion—synchronously photographed from five different points of view—in which one of the fore feet is thrust forward as far as it is possible for the horse to thrust it during any method of uniform progressive action.

A vertical line dropped from the nose of the animal in any one of these simultaneous photographs will intersect the leg much nearer to the fetlock or the pastern joint than to the knee.

The five figures are entirely free from any outlining or retouching.

In literature, the ancient poets and other authors—according to their translators—seem rarely to have made use of a distinctive name in their references to the rapid movements of animals They apparently preferred to indicate velocity of motion by some simile, or a comparison with the phenomena of nature.

These similitudes abound in Homer. A few selections, without reference to line or book, are taken from the "Iliad" as translated by Pope.

Should they interest the reader who is not already familiar with them, he will do well to read the poet from beginning to end, giving particular attention to the twenty-third book, describing the chariot-race at the funeral rites of Patroclus.

"High on his car he shakes the flowing reins.
His fiery coursers thunder o'er the plains."

"And now both heroes mount the glittering car·
The bounding coursers rush amidst the war."

"Saturnia lends the lash; the coursers fly:
Smooth glides the chariot through the liquid sky."

"The coursers fly before Ulysses' bow
Swift as the wind, and white as winter snow."

"He said; the driver whirls his lengthful thong:
The horses fly, the chariot smokes along"

"He lends the lash: the steeds with sounding feet
Shake the dry field, and thunder towards the fleet."

"The affrighted steeds, their dying lords cast down,
Scour o'er the fields, and stretch to reach the town."

"High o'er his head the circling lash he wields:
The bounding coursers scarcely touch the fields."

In the "Odyssey" we find—

"Ranged in a line the ready racers stand,
Start from the goal, and vanish o'er the strand.
Swift as on wings of wind, upborne they fly,
And drifts of rising dust involve the sky."

Xenophon (Spelman), in the "Expedition of Cyrus," says, "Patagyas . . . was seen riding towards them full speed;" and in the "Institution of Cyrus" the horses of the messengers are said to "fly swifter than cranes," which he, however, doubts.

Cæsar (Clarke), in his commentaries, "Wars in Gaul," writes that "Considius came galloping back;" and in the "Civil War" that "Pompey . . . rode full speed to Larissa."

Phaer, in the sixteenth century, causes Virgil, in "Æneidos," to say—

"In armour iointly ryde, hie shoutes vprise and clustring strakes
They gallop, and vnder their trampling feete the ground with breaking quakes."

Dryden renders the passage—

"The neighing coursers answer to the sound
And shake with horny hoofs the solid ground."

Pliny (Holland), describing a lion hunt, says, "Then he [the lion] skuds away, then he runneth amaine for his life."

Lucan (Rowe),—

"The fliers now a doubtful flight maintain
While the fleet horse in squadrons scour the plain."

Plutarch (North): "Hannibal . . . commanded the horsemen . . . to scurry to the trenches."

Tacitus (Murphy) speaks of "a Numidian horseman, posting at full speed."

The translators of these books used, of course, the phraseology that occurred to them as best indicating the, perhaps indefinite, motion expressed by the authors.

Coming to a more recent period, it is interesting to note the different words and expressions used by English authors during the last few centuries to denote extreme speed.

In a thirteenth-century manuscript, "Amis and Amiloun," of the Auchinlech Collections—

"On palfray, and on stede
He pryked both nyght and day
Till he come to his contrày,
There he was lord in dede."

Pricking was by early English writers used as synonymous with rapid speed.

Chaucer constantly makes use of the term. In "The Tale of Sir Thopas," Fyt I.—

" . . . priked as he were wood,
His faire steede in his prikynge
So swette, that men might him wrynge
His sydes were al blood."

In the fourteenth century "walop" was occasionally used. In the reproduction of the manuscripts of "Morte Arthure," by the Early English Text Society, occurs—

"Swerdes swangene in two sweltand knyghtez
Lyes wyde opyne welterande one walopande stedez,"

and by the same society, in the "Romance of William of Palerne," of date 1350—

"Or he wiste, he was war of the white beres
Thei went a-wai a wallop as thei wod [mad] semed."

In "Merlin," an anonymous manuscript of 1450, appears—

"Than the Kynge rode formest hym-self a grete walop."

In English literature the earliest known use of the word "gallop" occurs in an anonymous manuscript of the fifteenth century—"King Alisaunder"—deposited in the Bodleian Library—

> "Knyghtis wollith on huntyng ride;
> The deor galopith by wodis side.
> He that can his time abyde,
> At his wille him schal bytyde."

It is singular that this primitive use of the word should refer to the *rota*-gallop of the deer.

Gower, in "Confessio Amantis," has—

> "This knight after the kynges wille
> With spore made his hors to gone,
> And to the toune he came anone."

In "Froissart's Cronycle," translated by Berners, fifteenth-sixteenth century, we find—

> "Styll he goloped forth right, tyll he came into Arthoyes."

Spenser, in the "Faerie Quene," used "prick" and "gallop" indifferently—

> "So as they traveild, lo! they gan espy
> An armed knight towards them gallop fast."

> ". . . And as they forward went,
> They spied a knight faire pricking on the plaine."

Sir Philip Sidney, in "Arcadia," uses "gallop" exclusively as indicative of speed "Seeing such streams of blood, as threatened a drowning life, we galloped toward them to part them"

Shakespeare repeatedly uses "gallop" and "galloping." In *Macbeth*, iv. 1, for example—

> "I did hear
> The galloping of horse: who was't came by?"

and frequently as a metaphor, as in *Titus Andronicus*, ii. 1—

> "As when the sun salutes the morn,
> And, having gilt the ocean with its beams,
> Gallops the zodiac in its glistening coach,
> And overlooks the highest peering hills."

Among other words used by Shakespeare to denote the extreme speed of an animal are, "runs," "spurs," "spur-post," "high-speed," "skirre," "helter-skelter," "flying," etc.

Beaumont and Fletcher, in "Knight of the Burning Pestle," use "galloped amain."

In "Don Quixote," Part II. xxi., Cervantes, according to Skelton, speaks of "large carreere."

Dryden, in "Palamon and Arcite"—

> ". . . spurred his fiery steed
> With goring rowels to provoke his speed."

Butler, "Hudibras," Part I. canto iii.—

> ". . . then ply'd
> With iron heel his courser's side,
> Conveying sympathetic speed
> From heel of Knight to heel of steed."

Addison, *Spectator*, 56, "saw . . . a milk-white steed . . . full stretch."

Le Sage (Smollett) says Gil Blas went off "at a round gallop."

Sterne, in "Tristam Shandy," takes "a good rattling gallop."

Thomson, in the "Castle of Indolence," Canto II. vi.—

"Pricked through the forest to dislodge his prey."

Scott, "Marmion," Canto I. iii.—

"A horseman darting from the crowd;"

and in the same poem, Canto V. ix.—

". . . straining on the tighten'd rein
Scours doubly swift o'er hill and plain;"

also VI. xv.—

"The steed along the drawbridge flies."

Among other words and expressions used by Scott as indicative of extreme speed are, "rode amain," "rushed," "plunged," "headlong course," "pricked," "spurred fast," "full career," "speedy gallop," "full gallop," "bolted," "shot ahead," "sweep," and so forth.

Wordsworth, Prelude X.—

". . . beat with thundering hoofs the level sand."

In "Christabel," Coleridge's—

". . . palfrey was as fleet as wind,
. . . they spurred amain."

Byron's wolf in "Mazeppa," xii, had a "long gallop." In "The Giaour," he asks—

"Who thundering comes on blackest steed,
With slacken'd bit and hoof of speed?"

and in "Lara," Canto II. xxiv.—

"And instant spurr'd him into panting speed."

In the "Nurse's Story," "Ingoldsby Legends"—

"A queer-looking horseman . . . puts spurs to his hack, makes a dash through the crowd, and is off in a crack."

Combe, in "Dr. Syntax," has—

"The jockies whipped; the horses ran."

Sheridan, in "A Trip to Scarborough," speaks of "a running horse."

Matthew Arnold, in "Balder Dead," describes how "Odin gallop'd . . . like a whirlwind."

In "How they brought the Good News from Ghent to Aix," Browning, in eight lines of the poem, uses "gallop" ten times. It is possible the poet could have explained his reason for so doing.

Macaulay, in "Battle of the Lake Regillus," xxv., sends—

> ". . Black Auster
> Like an arrow from the bow,"

and in the same poem (iii.)—

> ". . . wolves came with fierce gallop."

In "The Revolt of Islam," by Shelley—

> ". . . with reinless speed
> A black Tartarian horse of giant frame
> Comes tramping . . ."

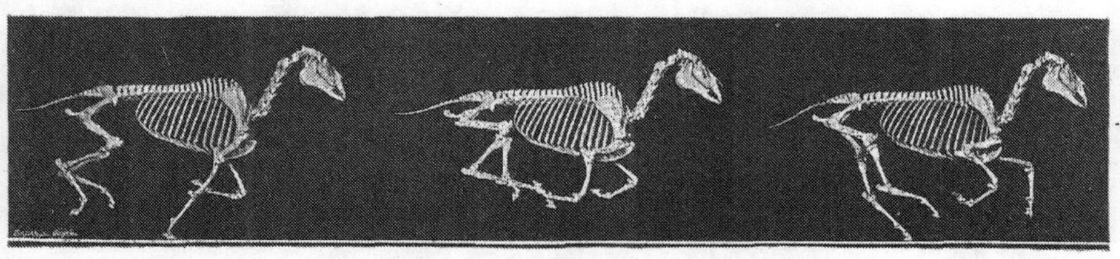

SOME PHASES IN THE GALLOP OF THE HORSE

Series 46.

TRANSVERSE-GALLOP.

ONE STRIDE IN TWENTY-ONE PHASES.

Thorough-bred Horse "Bouquet."

APPROXIMATE MEASUREMENTS

Foot-impacts:

78 inches. 1·95 metres	46 inches 1·15 metres.	90 inches. 2·25 metres.	60 inches. 1·50 metres.	Total length of stride 274 inches. 6·85 metres.

Time-intervals of phases · ·022 second. Distance-intervals 13⅜ inches.

Time of stride ·44 second. Strides to a mile · 231 Speed equivalent to a mile in 102 seconds

THE GALLOP.

TRANSVERSE-GALLOP.

Some consecutive phases of a representative stride by a thorough-bred horse while galloping at a speed of a mile in 102 seconds, or about 35 miles an hour.

 SERIES 47

TRANSVERSE-GALLOP.

ONE STRIDE

Thorough-bred Mare "Annie"

APPROXIMATE MEASUREMENTS.

Foot-impacts:

					Total length of Stride.
	88 inches. 2 20 metres.	40 inches. 1 00 metre.	84 inches. 2·10 metres.	60 inches. 1 55 metres	272 inches 6 85 metres.

Time-intervals 031 second. Time of stride: ·46 second Strides to a mile: 233 Speed equivalent to a mile in 107 seconds

Series 48.

TRANSVERSE-GALLOP.

An Incomplete Stride, Photographed Synchronously from Two Points of View.

Thorough-bred Horse "Bouquet."

APPROXIMATE MEASUREMENTS.

Foot-impacts:

56 inches.	46 inches.	86 inches	36 inches.	Total length of stride 224 inches
1 40 metres.	1 15 metres	2 15 metres	·90 metre.	5 60 metres.

Time-intervals: ·044 second

Series [illegible]

TRANSVERSE-GALLOP.

AN INCOMPLETE STRIDE, PHOTOGRAPHED SYNCHRONOUSLY FROM TWO POINTS OF VIEW.

Thorough-bred Horse "Bouquet"

APPROXIMATE MEASUREMENTS

Foot-impacts

				Total length of Stride
60 inches.	34 inches.	86 inches.	50 inches.	230 inches.
1·40 metres	·85 metre	2·15 metres.	1·25 metres.	5·75 metres.

Time-intervals · 037 second.

 SERIES 50.

TRANSVERSE-GALLOP.

ONE STRIDE, PHOTOGRAPHED SYNCHRONOUSLY FROM TWO POINTS OF VIEW.

Mare "Pandora."

Time-interval: ·042 second.

 SERIES 51.

TRANSVERSE-GALLOP.

ONE STRIDE IN EIGHT PHASES, PHOTOGRAPHED SYNCHRONOUSLY FROM TWO POINTS OF VIEW

Horse "Hansel"

Time-intervals· ·095 second

Series 52

TRANSVERSE-GALLOP.

One Stride.

The Buffalo, or Bison

Time-intervals .029 second.

Series 53

TRANSVERSE-GALLOP.

One Stride in Sixteen Phases.

The Goat.

Length of stride: 68 inches (1·70 metres) Time-intervals: ·029 second Approximate time of stride: ·43 second

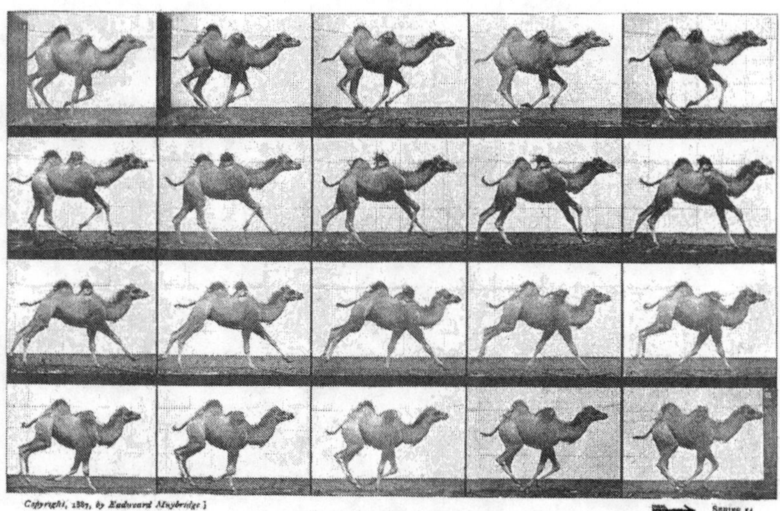

Series 54

TRANSVERSE-GALLOP

ONE STRIDE IN EIGHTEEN PHASES

The Bactrian Camel

Length of stride 140 inches (3·50 metres) Time-intervals ·032 second Approximate time of stride ·51 second

THE GALLOP.

Series 55.

TRANSVERSE-GALLOP.

AN IRREGULAR STRIDE IN NINE PHASES

The Cat

Time-intervals. 035 second

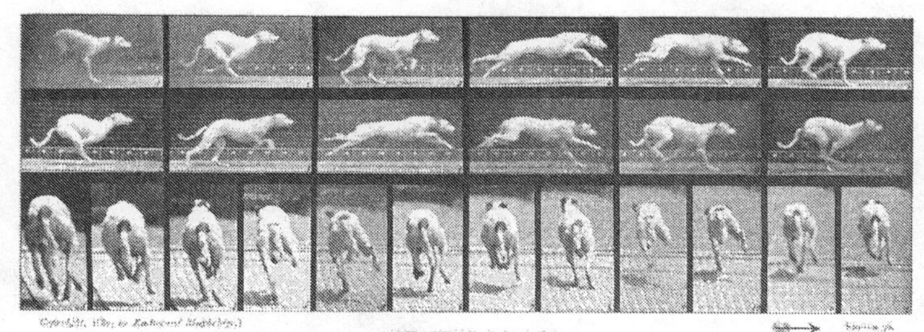

ROTATORY-GALLOP.

ONE STRIDE IN SIX PHASES

The Dog (racing hound)

Height, shoulder to ground 16 inches (40 metre) Length of stride: 114 inches (2.85 metres).

Time-intervals 049 second Approximate time of stride. ·25 second, equivalent to a mile in 139 seconds.

Series 57

ROTATORY-GALLOP.

One Stride, Photographed Synchronously from Two Points of View

Dog (Mastiff)

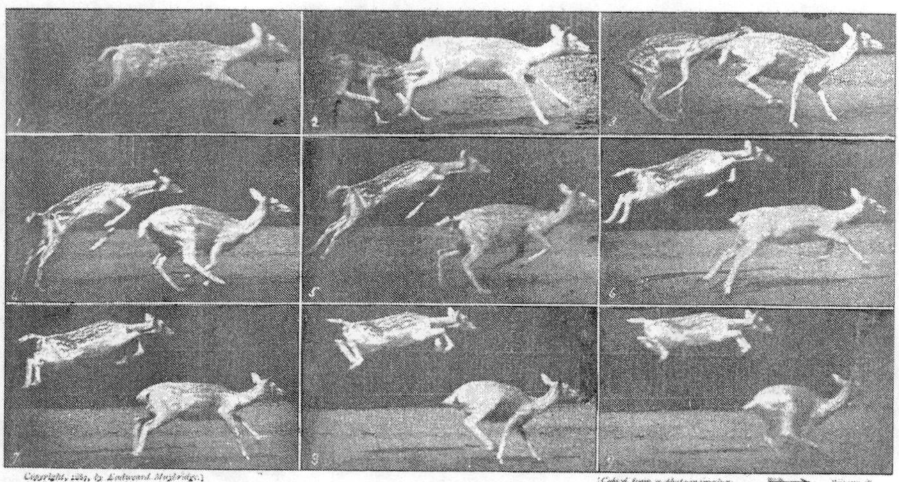

 [Copied from a photoengraving → Series 56

ROTATORY-GALLOP.

One Stride in Six Phases, and an Incomplete Leap of the Fawn

A Fallow Deer followed by its young Fawn

Series 59.

ROTATORY-GALLOP

ONE STRIDE IN FIFTEEN PHASES

The Wapiti, or Elk

Length of stride · 172 inches (4·35 metres). Time-intervals ·027 second Approximate time of stride: ·40 second

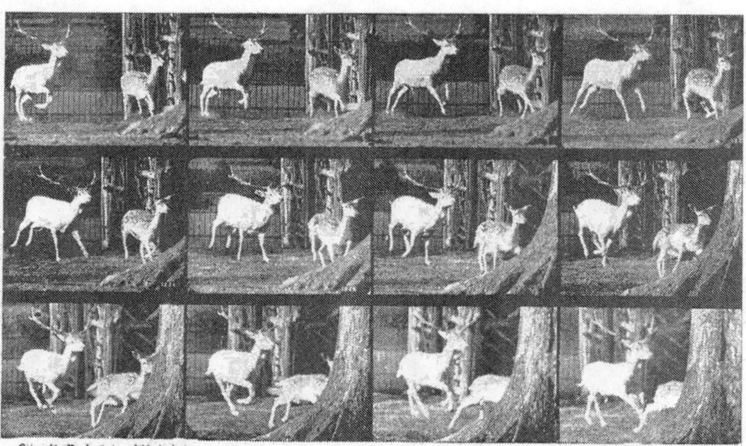

Series 60

ROTATORY-GALLOP

One Stride in Ten Phases.

The Fallow Deer

Time-intervals .052 second.

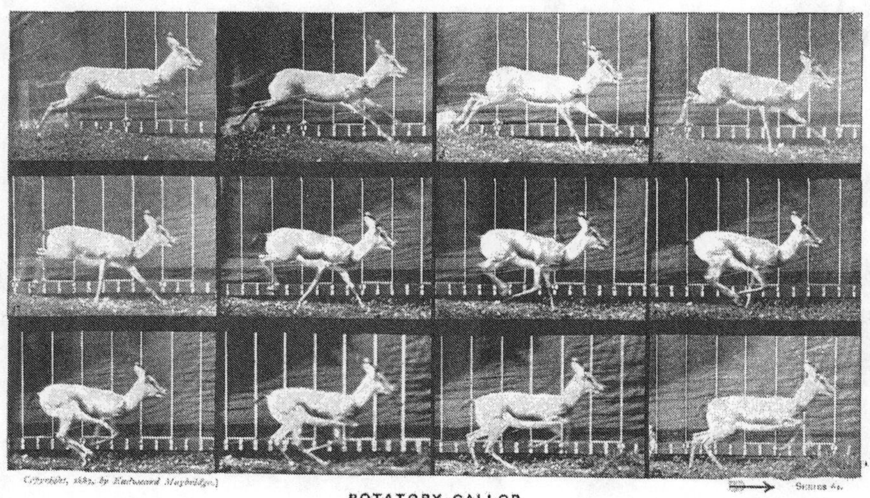

 Series 61.

ROTATORY-GALLOP.

AN INCOMPLETE STRIDE.

The Antelope

Length of stride 72 inches (1·80 metres)

THE GALLOP.

PHASES IN THE GALLOP OF THE CAT, CAMEL, DOG, RACCOON, WAPITI, BEAR, GOAT, BUFFALO, AND HOG

THE GALLOP.

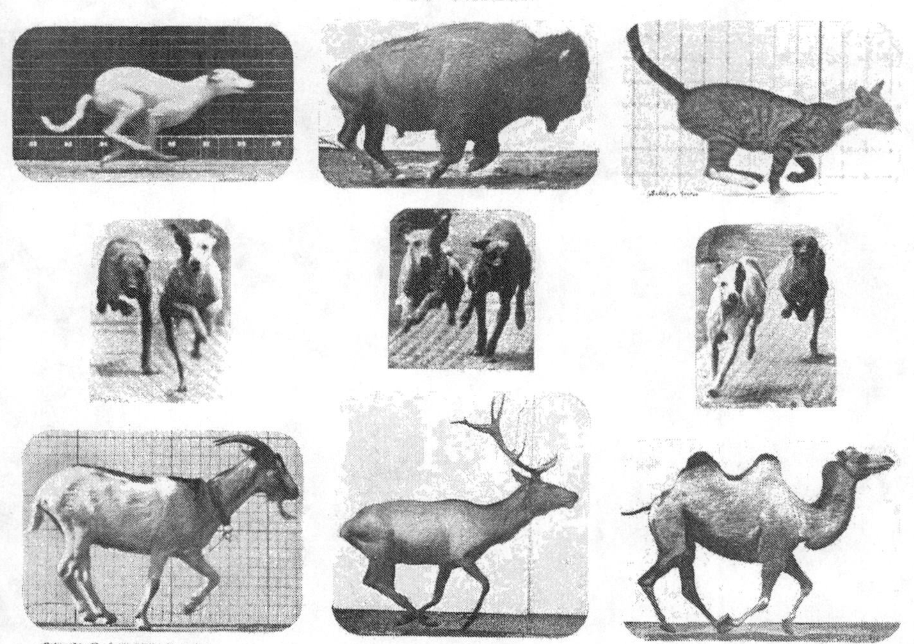

PHASES IN THE GALLOP OF THE DOG, BUFFALO, CAT, GOAT, WAPITI, AND CAMEL.

THE RICOCHET.

It is a curious fact that no one word in the English language has hitherto been applied to the system of progressive motion adopted by that class of animal, of which the kangaroo is the best known representative.

When we speak of the "walk" or the "gallop" of a horse we immediately associate with it the precise movement of the animal to which the word refers. The rapid motion of the wallaby, or the kangaroo, however, we have been accustomed to recognize only as a *series* or a *succession* of "bounds," "hops," "leaps," "jumps," or "skips;" this is the Australian practice of describing the movement. As no one of the words quoted suggests of itself the idea of continuous progress, and it being desirable that this system of locomotion should have some definite name, the author has drawn upon the military vocabulary for one. The word "ricochet" has long been in use by artillerists as a name for the skipping or bounding action of a projectile over the surface of the land or the water, and there seems no good reason why it should not be equally applicable to the skipping or bounding action of the kangaroo. It is preferable to employ an already well-recognized word for a similar movement—although of an inanimate object—than either to construct a new one, or to continue to use the combination of words hitherto necessary for distinguishing the motion. In this instance it would perhaps be advisable to Anglicize the word and give it the phonetic spelling of "rikosha."

It is the most simple of all methods of quadrupedal progress, and may be diagrammatically represented as—

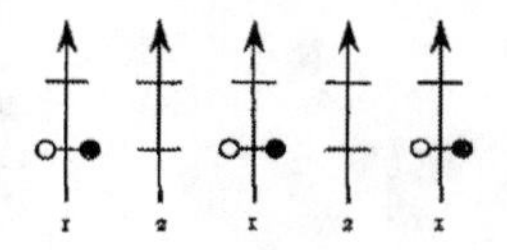

The action of the powerful tail of the kangaroo is not here recorded. It is, however, an important factor in its motion, it being brought in contact with the ground nearly simultaneously with the heel, and then effectively used to assist in the propulsion through the air.

Series 62 illustrates one stride of the ricochet made by a large-sized animal; and series 63 the commencement of a stride from a progress on all the four feet; to which latter movement it may be convenient, if not strictly correct, to apply the term "walk."

THE RICOCHET.

ONE STRIDE NEARLY COMPLETED.

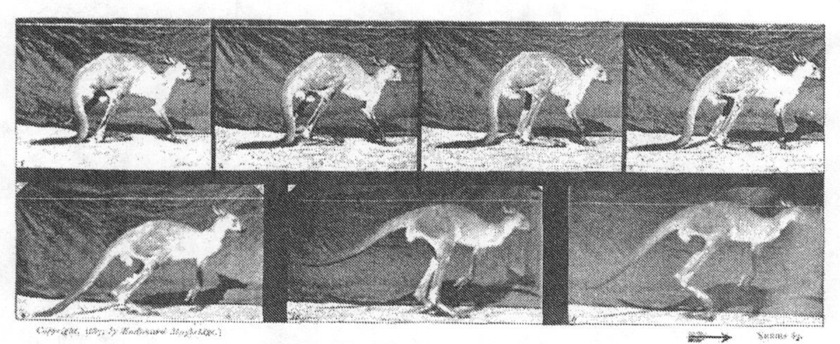

THE COMMENCEMENT OF A STRIDE

The Kangaroo

THE LEAP.

No inflexible rule can be laid down as to the details of a leap by any animal. The gathering together of the feet, the spring, the clearance of the obstacle, and the descent are contingent upon such a variety of circumstances that no law can be formulated of universal application. Investigation has proved that the same horse, with the same rider, jumping the same hurdle, under the same apparent conditions, will in two successive jumps present an entirely different series of phases. Take, for example, series 64. The last projecting force is shown in phase 7 to have been made by ●; but, contrary to the usual practice, the first contact in the descent is made on the lateral fore, which is also the last anterior to leave the ground in the preparation for the rise.

This peculiarity is repeated in series 65, the last and first contacts being with the same laterals. In series 66 the final spring is made from ○, and the first contact with ▲; in series 67 the last contact is with ●, and in the descent △ is the first to touch the ground. In these two latter seriates we find the general practice of horses in leaping. The first contact with the ground usually takes place on the foot diagonal to that which effected the final projecting force.

Had these four leaps been made by an untrained animal it might be supposed the irregularities proceeded from a lack of experience; but Pandora was a mare of national reputation, and had frequently, with a hundred-and-eighty-pound man on her back, cleared, without touching, a stone fence five feet high.

The particular hind-foot with which an animal, when clearing an obstacle, will give its ultimate propulsion is a matter of convenience, depending entirely upon the relative positions of the feet when the gathering for the leap, at the calculated distance, is decided on.

Seriates 68 and 69 illustrate parts of two jumps by

a good hunting-horse, but of much less capability than Pandora. They demonstrate the general preparations for a leap, and the renewal of regular motion after it is executed.

The jump of a cat is subject to a far greater number of varying conditions than those which attend a similar movement by the horse.

The comparative time-intervals of the selected phases on p. 225 can be ascertained by a reference to their respective numbers.

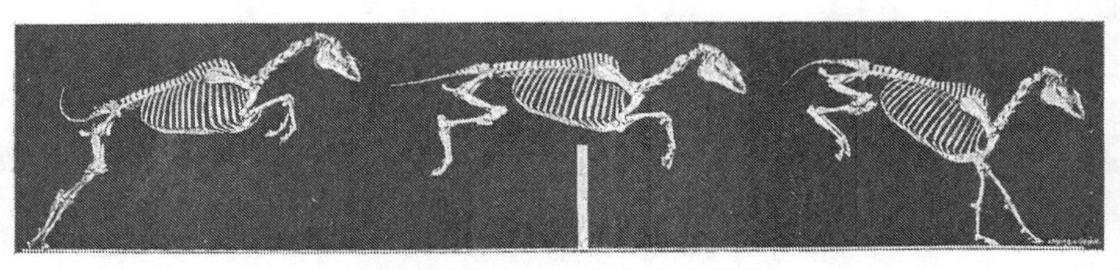

SOME PHASES IN THE LEAP OF A HORSE

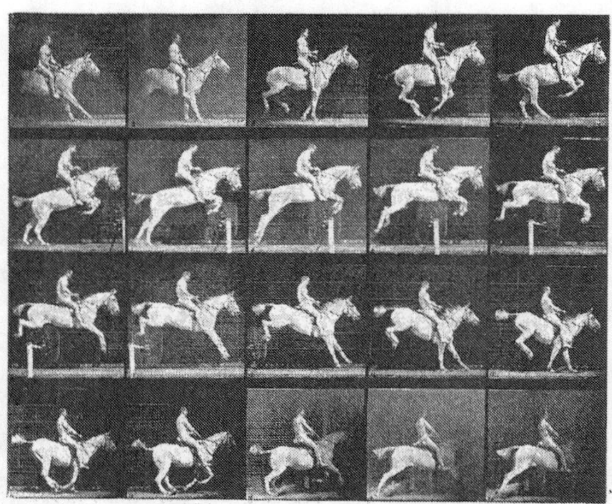

 Series 64

OVER A BAR ONE YARD HIGH.

Mare "Pandora"

Springing from ●, landing on ▲

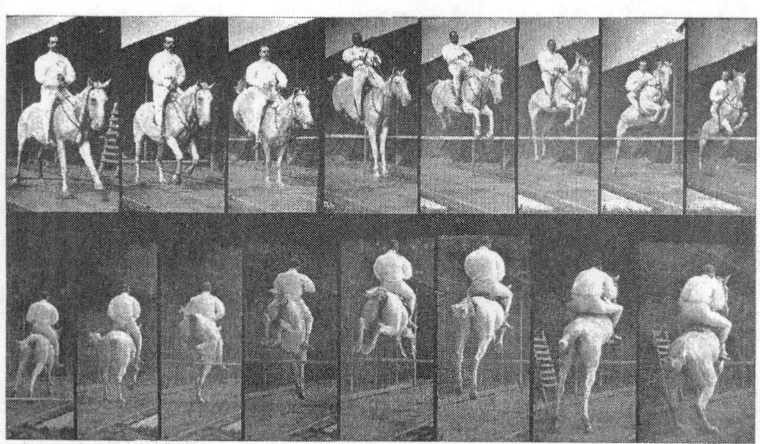

 ← Series 65

OVER A BAR ONE YARD HIGH

Photographed Synchronously from Two Points of View.

Mare "Pandora"

Time-intervals: 1/49 seconds. Springing from ◆, landing on ▲.

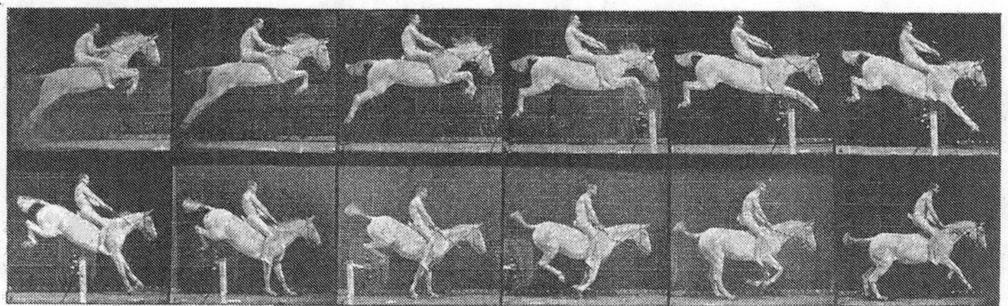

Series 66.

WITHOUT A SADDLE OVER A BAR ONE YARD HIGH

Mare "Pandora"

Time-intervals: 60 second. Springing from ○, landing on ▲

Series 67

WITHOUT A SADDLE OVER A BAR ONE YARD HIGH.

PHOTOGRAPHED SYNCHRONOUSLY FROM TWO POINTS OF VIEW

Mare "Pandora"

Time-intervals: 120 second Springing from ●, landing on △.

THE LEAP.

Series 66

PREPARATIONS AND COMMENCEMENT.

Horse "Daisy"

Springing from O

THE LEAP.

 SERIES 69.

CLEARANCE, LANDING, AND RECOVERY

Horse "Daisy."

Springing from ●, landing on △.

THE LEAP.

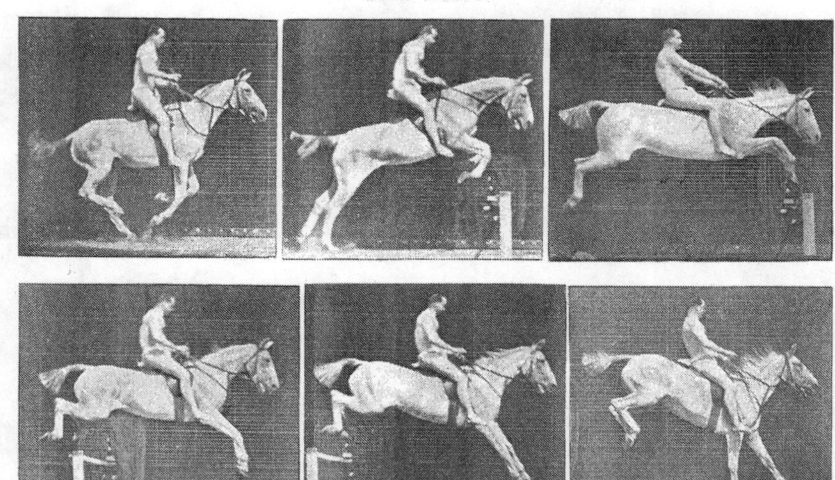

LATERAL PHASES OF SOME LEAPS.

Mare "Pandora."

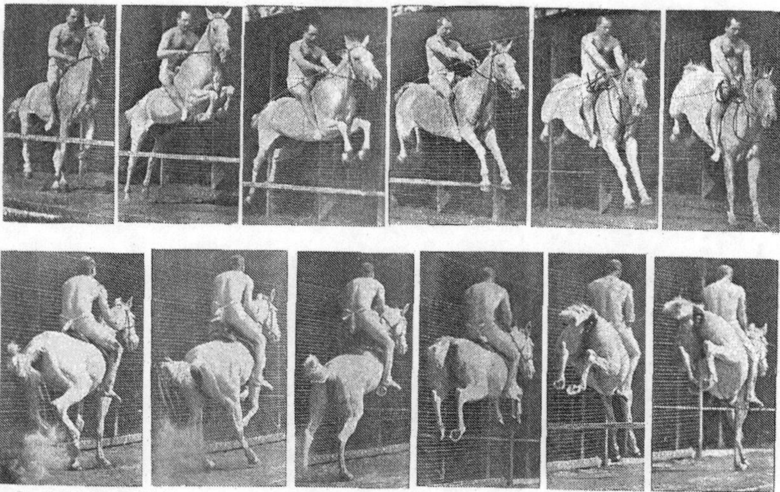

FORE-SHORTENED PHASES OF SOME LEAPS.

Mare "Pandora"

 [Enlarged from a photo engraving

A CAT FRIGHTENED WHILE TROTTING

Selected phases from a series of which the time-intervals were recorded ·035 second.

THE BUCK AND THE KICK.

It being difficult to obtain a horse who was sufficiently amenable to discipline as to buck and kick at the word of command, resort was had to a circus mule, who had undergone a regular course of instruction in those accomplishments. Seriates 70 and 71 represent two different actions, arranged on one page to facilitate comparison. The first two lines give a very fair illustration of a buck, followed by a one-legged kick; the following two lines admirably realize the caprices of a high-kicker. Phase 4 of the second series very faithfully reproduces in life a carved slab of a wild ass in the Assyrian department of the British Museum.

THE BUCK AND THE KICK.

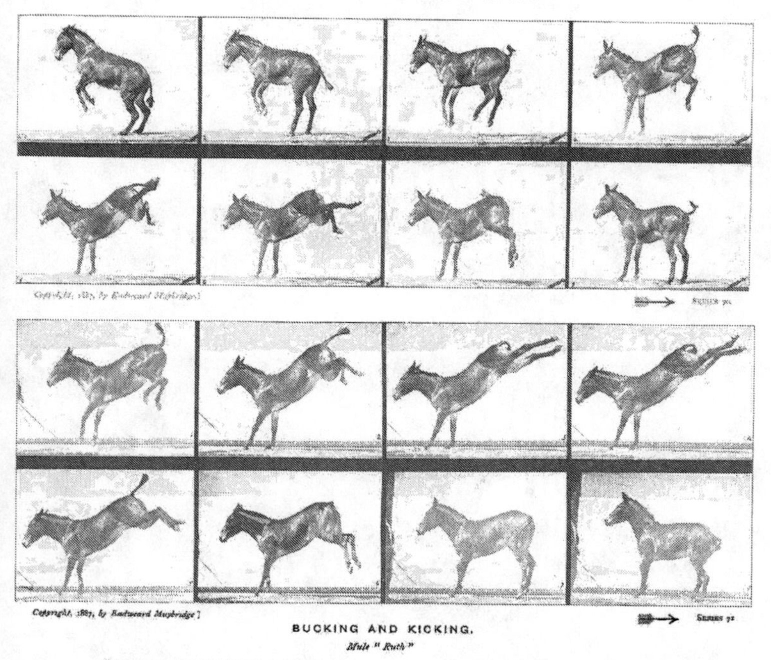

BUCKING AND KICKING.

Mule "Ruth"

Time-intervals of upper two lines: ·091 second. Time-intervals of lower two lines ·065 second

THE BUCK AND THE KICK.

SOME PHASES OF BUCKING AND KICKING

Mule "Ruth."

CHANGE OF GAIT.

The precise methods by which changes from the walk to the trot, or from the trot to the gallop, are effected, have always been disputable subjects. The former change is demonstrated in series 72, the latter in series 73. In each of these the horses were dragging a racing sulky, which has been obliterated to facilitate examination; with the same object in view dots have been made in proximity to the right pasterns.

Series 74 is a change from the rack to the gallop, 75 illustrates some phases attained by a thorough-bred Kentucky mare, after landing from a jump of twenty-six feet three inches, by actual measurement, over a bar one and a half yards high. Phases 23 and 24 are interesting, and suggest the Byzantine treatment of the gallop.

All of these changes were accidental, and therefore perfectly natural; 72, 73, and 75 have been filled in to make them more distinct for study, the exact outlines having been carefully followed. Series 74 remains in the precise condition of the original negative, with all its faults.

These four seriates were photographed in 1879.

CHANGE OF GAIT.

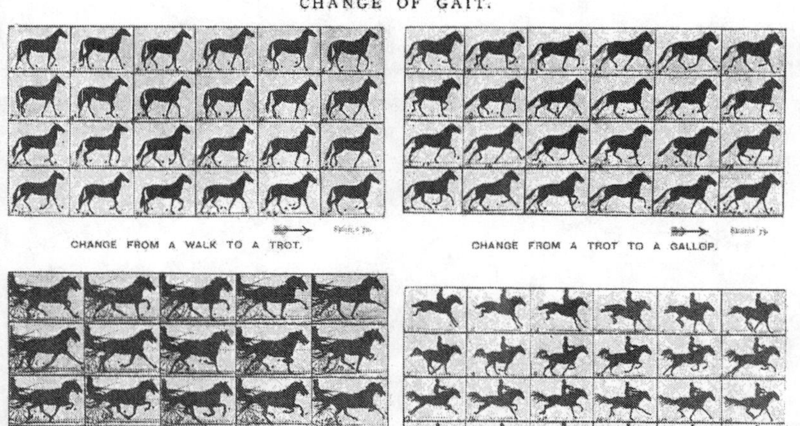

CHANGE FROM A WALK TO A TROT.

CHANGE FROM A TROT TO A GALLOP.

CHANGE FROM A RACK TO A GALLOP

RECOVERY AFTER A LEAP OF EIGHT AND THREE-QUARTER YARDS IN LENGTH OVER A BAR ONE AND A HALF YARDS HIGH.

Photographed at Palo Alto 1879

The right feet of the horses are indicated by dots near the pasterns

THE FLIGHT OF BIRDS.

PIGEONS.

Photographed at Palo Alto, 1879.

THE subjects of flight and soaring present so many intricate problems that the author is reluctantly compelled to relinquish his attempt to elucidate them. His investigation, however, brought to light some facts which, although they had been theorized upon, had never been proved. Phases 5, 6, 7, 16, 17, and 18 of the cockatoo, series 79, demonstrate that the primary feathers of a bird's wing, although interlocked in the downward stroke, are separated, and their thin edges turned in the direction of their movement during the recovery. This partial revolution of the primary feathers is also distinctly seen in the pigeon, seriates 76 and 77; the cockatoo during a second flight, series 80, the vulture, series 78; and it probably occurs in the flight of all birds, large or small. The dissection of a crane, made at the author's request, proved the combined movement of the primary feathers to be under the control of the bird, and independent of any motion of the wing itself. Whether the act of soaring is accomplished by some, hitherto imperceptible, motion of the primary feathers may perhaps engage the attention of future investigators.

 Series 76.

ONE FLAP OF THE WINGS IN SEVEN PHASES

PHOTOGRAPHED SYNCHRONOUSLY FROM TWO POINTS OF VIEW

Homing Pigeon.

Time-intervals: ·019 second

Copyright, 1887, by Eadweard Muybridge] Series 77

ONE FLAP OF THE WINGS IN SEVEN PHASES.

PHOTOGRAPHED SYNCHRONOUSLY FROM TWO POINTS OF VIEW.

Homing Pigeon

Time-intervals ·020 second

Copyright, 1887, by Eadweard Muybridge]

Series 78

ONE FLAP OF THE WINGS IN FIFTEEN PHASES.

Vulture

Time-intervals. ·019 second.

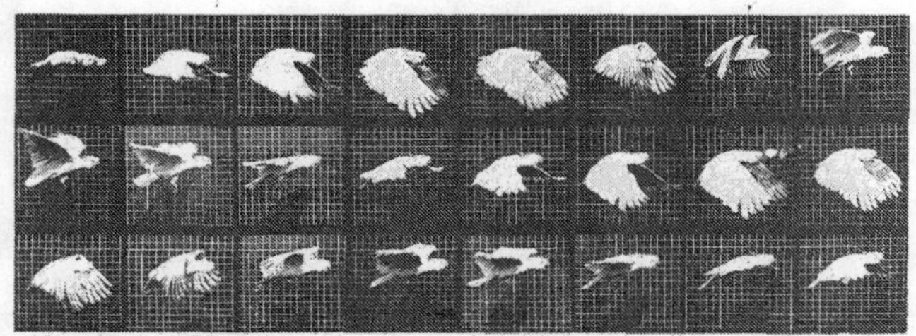

 Series 79

TWO FLAPS OF THE WINGS

The Sulphur-Crested Cockatoo.

Time-intervals: ·017 second.

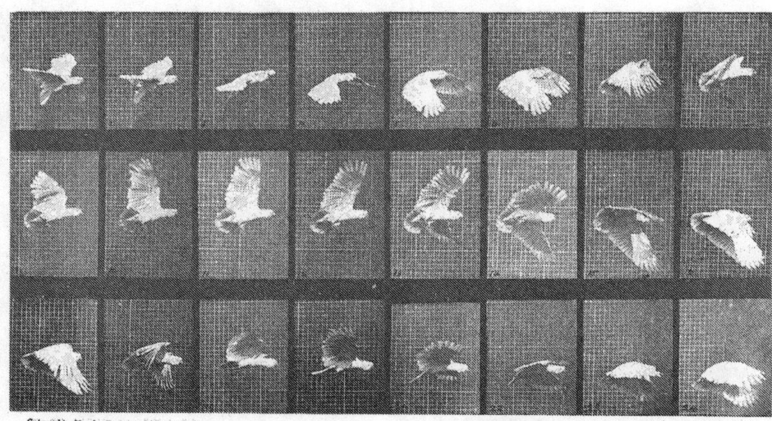

Copyright, 1887, by Eadweard Muybridge] Sabnis Sc.

AN IRREGULAR FLAP OF THE WINGS IN TWELVE PHASES.

The Sulphur-Crested Cockatoo

Time-intervals: ·028 second.

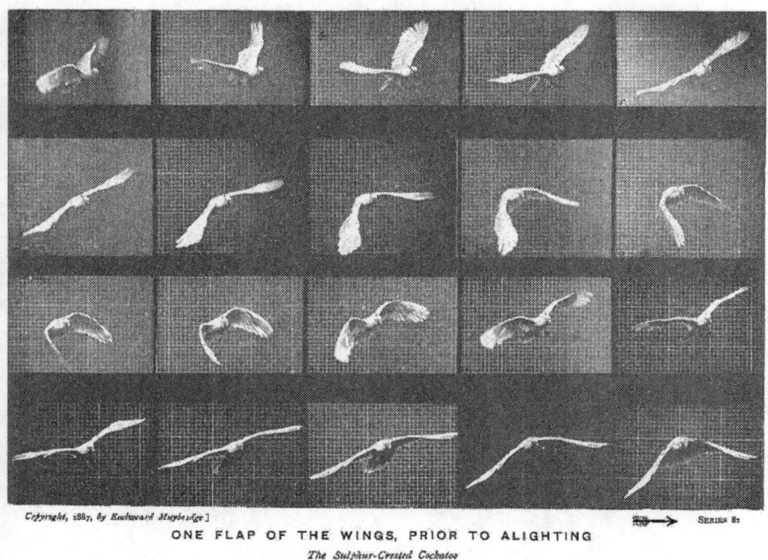

 SERIES 81

ONE FLAP OF THE WINGS, PRIOR TO ALIGHTING

The Sulphur-Crested Cockatoo

Time-intervals . 016 second

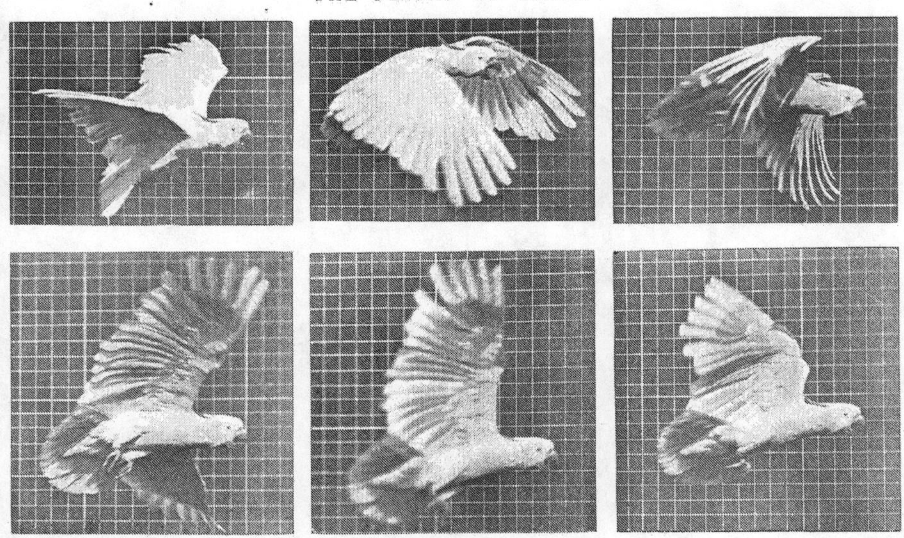

SOME PHASES IN THE FLIGHT OF A COCKATOO

THE FLIGHT AND WALK OF BIRDS.

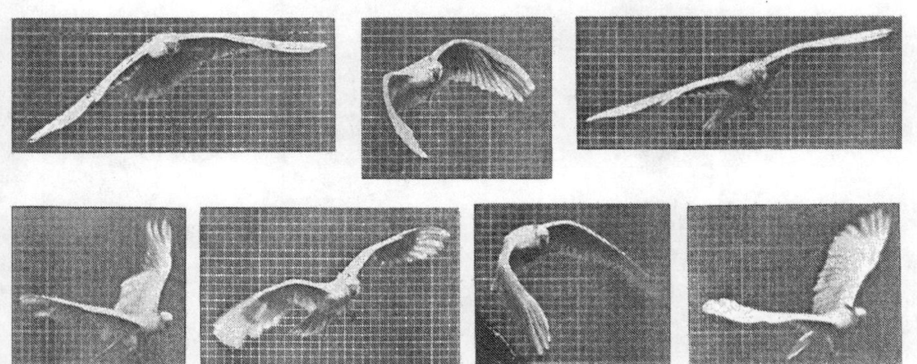

SOME PHASES IN THE FLIGHT OF A COCKATOO.

SOME PHASES IN THE WALK OF AN ADJUTANT

 SERIES 80

WING MOVEMENTS WHILE RUNNING

A HALF-STRIDE IN EIGHT PHASES, PHOTOGRAPHED SYNCHRONOUSLY FROM TWO POINTS OF VIEW

The Ostrich

RECORDS OF MOVEMENTS FROM OBSERVATION.

THE illustrations in this work include representatives of all the animals, the movements of which were photographically investigated by the writer.

The analyses of these movements being demonstrated facts, are not open to controversy. It would have been desirable, however, to have photographed many of the animals while they were enjoying more freedom of movement than that afforded by the gardens of a Zoological Society, but the difficulties attending a satisfactory investigation under their natural conditions of life were, at the time, too great to be surmounted.

It may, therefore, be desirable to include a few quotations from some well-known naturalists who have had the opportunity of observing the movements of wild animals in their natural haunts.

THE ELEPHANT. Sir Samuel W. Baker, in "Wild Beasts and their Ways," chap. ii., says—

"I consider that the African elephant is capable of a speed of fifteen miles an hour, which it could keep up for two or three hundred yards, after which it would travel at about ten miles an hour, and actually accomplish the distance within that period. The Asiatic elephant might likewise attain a speed of fifteen miles for perhaps a couple of hundred yards, but it would not travel far at a greater pace than eight miles an hour."

(Chap. iii.) "Although an elephant is capable of great speed it cannot jump, neither can it lift all four legs off the ground at the same time; this peculiarity renders it impossible to cross any ditch with hard perpendicular sides that will not crumble or yield to pressure, if such a ditch should be wider than the limit of the animal's extreme pace. If the limit of a pace should be 6 feet, a 7-foot ditch would effectually stop an elephant."

It has already been suggested that the elephant has two gaits only, the walk and the amble. Baker's experiences confirm the opinion that the animal is incapable either of trotting, racking, cantering, or galloping. The correctness of Sir Samuel's observations have been endorsed, in a letter to the writer, by Mr. Frederick C. Selous.

THE RHINOCEROS. In the same work, chap. xvi., Baker writes—

"When the vast bulk of a rhinoceros is considered, it is astonishing to see the speed that this heavy animal can attain, and continue for a great distance. I have hunted them in company with the Arabs, and for at least 2 miles our horses have been doing their best, keeping a position within 5 or 6 yards of the hind quarters, but, nevertheless, unable to overtake them before they reached an impenetrable jungle. It is the peculiar formation of the hind legs which enables the rhinoceros to attain this speed; the length from the thigh to the hock is so great that it affords immense springing capacity, and the animal bounds along the surface like a horse in full gallop, without the slightest appearance of weight or clumsiness."

Of the same animal, Selous, in "Travel and Adventure in South-East Africa," chap. xxv., says—

"A black rhinoceros trotted out into the open, having no doubt got my wind as I passed. . . . He had broken from a trot into a gallop before I fired; but on receiving the shot went a good deal faster, at the same time snorting violently. . . . A black rhinoceros can gallop at an extraordinary pace for so heavy a beast; indeed, it is just as much as a good horse can do to overtake one, so that as I ranged alongside, my horse, a powerful stallion, was going at his utmost speed."

Baker and Selous are probably the two best authorities on the actions of African wild beasts in their native haunts, and they agree in their observations of the extraordinary speed the rhinoceros is capable of attaining, rivalling, apparently, the fastest motion of an elephant. The description of the fast gait of the animal by these celebrated hunters does not correspond with that of the amble; Baker compares it to the "full gallop" of a horse, and Selous says it broke "from a trot into a gallop."

Neither of these keen observers would be likely to mistake the motion of the gallop or the canter for the more steady and uniform progress of an animal when trotting or ambling.

It is very desirable that some African explorer should succeed in obtaining photographs of the rhinoceros under full speed, as, like the hippopotamus, it will perhaps in a few more years be exterminated. A single lateral exposure will, under favourable conditions, be quite sufficient to determine the character of the movement.

THE HIPPOPOTAMUS. The walk of the hippopotamus, according to the observation of the writer, conforms to the law governing that of purely terrestrial vertebrates. In "Wild Beasts and their Ways," chap. xii., Baker gives an interesting account of the speed of the animal when entirely submerged—

"A hippopotamus can move at a considerable pace along a river's bed. We had proof of this while running down the Bahr Giraffe with the steamer, the speed with the stream being about 10 knots an hour. . . . It was some time before we actually gained upon it, but when the engineer put on full steam, there could be no doubt of our superiority in speed."

While under water it is probable that the hippopotamus can trot, and with a long stride make considerable progress along the bed of the river without the actual support of its legs. On dry land it is hardly probable that its fastest gait can be other than the amble; possibly a trot, but with a very brief period, if any, of non-support.

THE GIRAFFE. Unfortunately no giraffe was available for the writer's investigation. Selous, in "Travels and Adventures in Africa," chap. xxvi., says—

"There were sixteen of these stately beasts in all, and a grand sight it was to view so many of them together. They . . . allowed us to approach to within two hundred yards of them before starting off at their peculiar gallop. (N.B.—Giraffes never trot, as they are so often represented to do in drawings. They have but two paces, a walk and a gallop or canter, and break at once from one into the other)."

In his interesting book, Selous has a picture of a giraffe walking; the animal is supported on the left laterals, the right hind-foot is approaching the place from which, in close proximity thereto, the fore-foot has just been lifted —a phase somewhat like 11 of the camel, series 13; or 4 of the horse, series 3. The same characteristic has been observed by the writer in the walk of a giraffe.

Baker, in "Albert Nyanza Great Basin of the Nile," chap. viii., describing a giraffe hunt, says—

"A good horse is required, as, although the gait of a giraffe appears excessively awkward from the fact of his moving the fore and hind legs of one side simultaneously, he attains a great pace, owing to the length of his stride, and his bounding trot is more than a match for any but a superior horse."

In the same book is a picture of Sir Samuel himself pursuing a herd of giraffes; the animals are represented as racking, the phase selected being similar to 7 of series 42.

In "Wild Beasts and their Ways," chap. xix., Baker again alludes to the fast pace of a giraffe—

"It moves like a camel, both legs upon the same side simultaneously. The long neck swings ungracefully when the animal is in rapid motion, and the clumsy half-canter produces the appearance of lameness"

The writer is inclined to believe that, when hard pressed, the rack of the giraffe, like that of the camel, will be exchanged for the transverse-gallop.

THE KANGAROO. Mr. W. Saville Kent, in his valuable work "The Naturalist in Australia," says of this animal—

"All the Macropodidæ are distinguished by the preponderating length of their hinder limbs, upon which alone they progress under any stimulus to rapid movement, by a characteristic series of leaps and bounds."

To this method of progress, it will have been seen, the writer has applied the name of "ricochet."

REPTILES. The motion of reptiles was not included in the photographic researches of the writer, but a few remarks founded on his observation of the use they make of their limbs may not be irrelevant.

An alligator, during an ordinarily slow walk on dry land, will move his feet in the same consecutive order, and with the same alternations of support, as a horse grazing in the fields. An acceleration of this pace results in the diagonal legs moving in pairs, much in the same order as those of a horse while trotting; whether, during a more rapid motion, the body of an alligator is unsupported for any portion of its stride was not determined, from inability to obtain a faster speed with the reptile experimented with.

The movements of the crocodiles, the lizards, and other reptiles of the same general formation, probably correspond with those of the alligator. The chameleon was carefully observed while walking on the ground, and while climbing the branch of a shrub. In both instances the movement of the limbs corresponded with the slow walk of an alligator.

The walk of the Gallapagos turtle and of the common garden tortoise, disclosed the fact of their bodies being supported on a pair of diagonals, alternately with three feet; the succession of foot-fallings conformed to the general law governing the same movement in other vertebrates.

In the "Naturalist in Australia," Kent gives a most interesting description of the peculiar motions of the chlamydosaurus, or the frilled-lizard. The Roebuck Bay specimens brought to England by him—

". . . were in vigorous health, and at the first trial when set at liberty, ran along almost perfectly erect with both their fore limbs and tail elevated clear of the ground.

'The distance the chlamydosaurus will traverse in this remarkably erect position may average as much as 40 or 50 feet at a stretch; when, after resting momentarily on its haunches, it starts off again. When, however, a short stretch of a few yards only has to be covered, the animal runs on all fours. . . Professor Huxley had no hesitation in assigning to this type an erect bipedal method of locomotion."

THE FLIGHT AND SOARING OF BIRDS. The attention of the writer was first directed to the soaring of birds during a southern tour of the United States early in the fifties, when he watched a buzzard wheeling around, at various elevations, for the space of an hour, without the slightest apparent effort of motion.

He once startled an eagle from a peak of the Sierra Nevada mountains; the bird gave two or three flaps of its wings, and without any further visible exertion, soared across the Yosemite Valley, and landed on another peak of the range, not less than three miles distant. The time was early in the morning, when there was not enough wind to extinguish a match struck in the open air; yet the time in which the bird traversed this distance was not more than a few minutes.

In "A Naturalist's Voyage," chap. ix., Darwin gives an interesting description of the soaring of the condor—

"When the condors are wheeling in a flock round and round any spot, their flight is beautiful. Except when rising from the ground, I do not recollect ever having seen one of these birds flap its wings Near Lima I watched several for nearly half-an-hour, without once taking off my eyes, they moved in large curves, sweeping in circles, descending and ascending without giving a single flap. As they glided close to my head I intently watched, from an oblique position, the outlines of the separate and great terminal feathers of each wing, and these separate feathers, if there had been the least vibratory movement, would have appeared as if blended together, but they were seen distinct against the blue sky. The head and neck were moved frequently, and apparently with force, and the extended wings seemed to form the fulcrum on which the movements of the neck, body, and tail acted. If the bird wished to descend, the wings were for a moment collapsed; and when again expanded with an altered inclination the momentum gained by the rapid descent seemed to urge the body upwards with the even and steady movement of a paper kite. In the case of any bird soaring, its motion must be sufficiently rapid, so that the action of the inclined surface of its body on the atmosphere may counterbalance its gravity. The force to keep up the momentum of a body moving in a horizontal plane in the air (in which there is so little friction) cannot be great, and this force is all that is wanted. The movement of the neck and body of the condor, we must suppose, is sufficient for this However this may be, it is truly wonderful and beautiful to see so great a bird, hour after hour, without any apparent exertion, wheeling and gliding over mountain and river"

The writer has frequently, while crossing the Atlantic, carefully watched with a powerful binocular glass, the motion of gulls while soaring quite close to and around the stern of a steamer, but notwithstanding the failure of his efforts, and those of others, to detect any motion in the primary feathers of the wings, he ventures the opinion that the power possessed by a bird of causing them to make a partial revolution, independently of any action of the wing itself, must be considered as a necessary factor in a solution of the problem of soaring.

APPENDIX.

In the Introduction, reference was made to the elaborate book on Horsemanship by the Marquis of Newcastle, originally published in the French language at Antwerp, 1658. The following extracts are taken from the Preface of the English edition of the work, published at London, 1743:—

"I might make an Article here regarding the Stile in which this Book is writ: But I think it sufficient to observe to my Readers, that I neither write as a Wit myself, nor for a Gentleman of Wit. Educated in the Stable, in the Stud in the Manage, in the midst of Horses in the Army, I have never been a Member of the French Academy. I write for those who, like myself, make it their Profession to be among Horses; it is enough that I make myself understood by them, by a proper Use of the Terms of Art, in which I presume I have pretty well succeeded."

A chapter in this book is devoted to "The Movements of a Horse in all his Natural Paces," which are described as follows:—

"The Walk.—A Horse in walking has two of his feet in the air, and two upon the ground, which move otherways at the same time, one fore and one hind-foot, which is the movement of a gentle trot.

"The Trot.—The action of his legs in this movement is two feet in the air, and two upon the ground, which he moves crossways at the same time; one fore and one hind-foot across, which is the movement of the walk for the movement of a horse's legs is the same in walking as in trotting, where he moves them cross-ways, two in the air across, and two upon the ground at the same time; so that those which were across in the air at one time, are afterwards in the same situation upon the ground, and so *vice versâ*. This is the real movement of a horse's legs in trotting.

"The Amble.—A horse in this action moves both legs on the same side; for example, he moves his two off-legs both before and behind at the same time, while those of the near side are at a stand; and when those two which were in motion before touch the ground, he moves the other side, *viz.* the fore and hind leg on the near side, and the off-legs are then at rest. Hence a pacing horse moves both legs on one side, and changes the side at each motion, having both legs on the same side in the air, and those of the other side upon the ground at the same time, which motion is the perfect amble.

"The Gallop.—Galloping is a different movement; for in this pace a horse can lead with which leg the rider pleases, but the leg on the same side must follow it; I mean when he gallops directly forward, and then this is a true gallop. But that the leading of the fore-leg may be rightly understood, which ought to be followed by the hind-leg of the same side, the leg moves in the following manner: for example, if the fore off-leg leads, it consequently follows by such leading, that the same fore-leg ought to be before the other fore-leg, and the hind-leg on the same side ought to follow, which hind-leg ought to be before the other hind-leg, which is the right gallop.

"But in order to understand it the better, the motion in galloping is in this manner: the horse raises his two fore-legs at the same time in the action I have described, which is one leg before the other, and when his fore-legs come down, before they touch the ground, they are immediately followed by those behind; so that, as I have said before, they are all in the air at the same time: for his hind-legs begin to move when the fore-legs begin to fall, by which the whole horse is entirely in the air. How would it otherwise be possible, that a horse in running should leap twice his length, if the motion of the gallop was not a leap forwards?

This description is very just both with respect to the motion and posture of a horse's legs in galloping, which, though it be true is not easily perceived in a gentle gallop, but very visible in a swift one, where the motion is violent: I say, his four legs may then plainly appear to be in the air at the same time, running being no more than a quick gallop, the motion and posture of a horse's legs being entirely the same.

"RUNNING.—The motion of a horse and the action of his legs are the same in running as in galloping, the different velocity of the motion only excepted, so that running may be properly called a swift gallop, and a gallop a slow running. This is the true movement in running. The trot is the foundation of a gallop; and the reason is, because the trot being crossways, and a gallop both legs on the same side, if you put a horse upon a trot beyond the speed of that pace, he is obliged when his off fore-leg is lifted up, to set down his near hind-leg so quickly, that it makes the hind-leg follow the fore-leg on the same side, which is a real gallop; and for this reason a trot is the foundation of a gallop.

"A gallop is the foundation of the Terre-à-terre, the motion of the horse's legs being the same. He leads with the fore-leg within the Volte, and the hind-leg on the same side follows. You keep him only a little more in hand in Terre-à-terre, that he may keep his time more regularly.

"I could wish that Pacing was excluded from the Manage, that action being only mixed and confused, by which a horse moves both legs on the same side, and shifts them each movement; and this is as directly contrary to the Manage as is possible, if, from an Amble you would put a horse to a gallop, for when he is upon a trot you may push him to a gallop, but being upon the amble you must stop him upon the hand before he can gallop."

A HORSE REARING.
Photographed at Palo Alto, 1879.

PRINTED BY WILLIAM CLOWES AND SONS, LIMITED, LONDON AND BECCLES

CPSIA information can be obtained
at www.ICGtesting.com
Printed in the USA
LVHW051835180523
747412LV00004B/62